Mathematics and Computing/Technology
A Third-level Course
MT365 Graphs, Networks and Design

GRAPHS

2

Prepared by the Course Team

TREES

Study guide

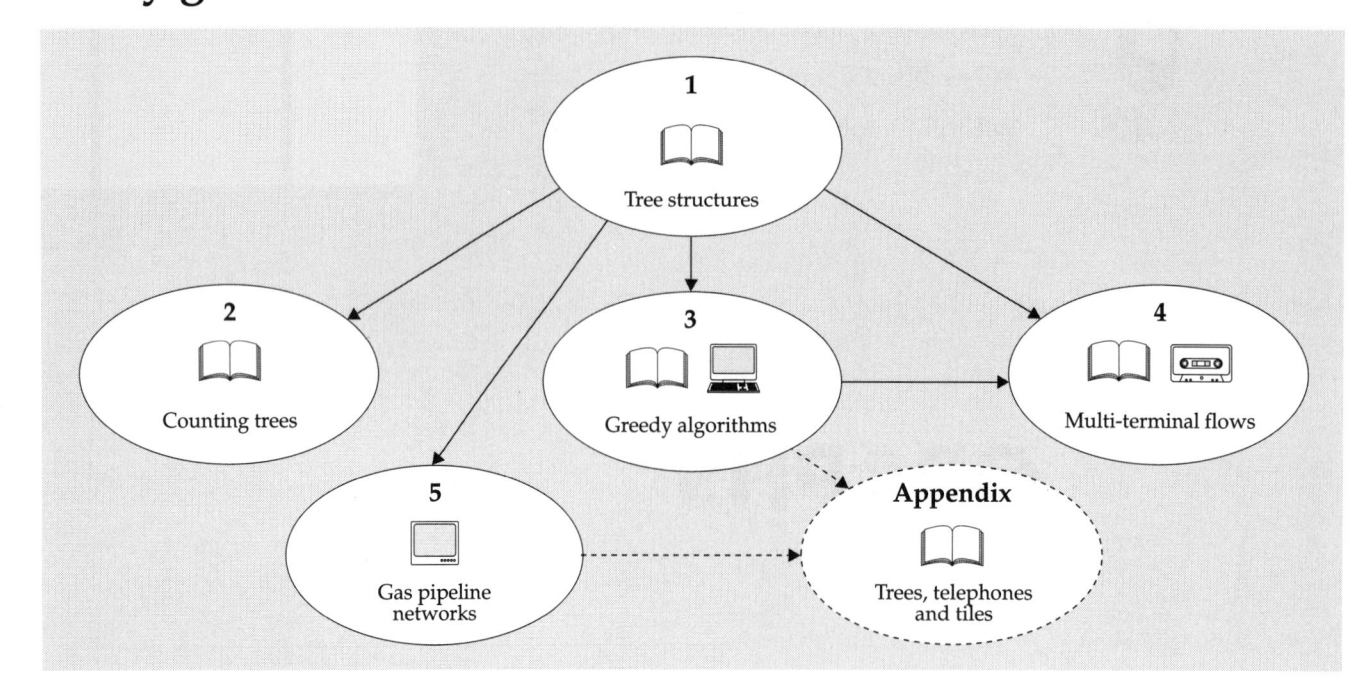

Section 1 revises and extends ideas introduced in *Graphs 1*, and investigates the mathematical properties of trees. Section 2 contains a subsection on counting binary trees, and you should not feel unduly discouraged if you have trouble with the mathematical details — the important thing is to try to understand the main ideas. Sections 3 and 4 are likely to take most of your time and effort. Section 3 revises algorithms you met in the *Introduction* unit. There are computer activities associated with Section 3, and an audio-tape sequence associated with Section 4. Section 5 is the television section; the television programme can be watched at any time during your study of the unit.

The Appendix is optional, and is included for interest only. It contains an article which discusses several of the topics in this unit and the *Introduction* unit. We suggest that you read it after you have completed your study of this unit.

The Open University, Walton Hall, Milton Keynes, MK7 6AA.

First published 1995. Reprinted 1997, 2000

Copyright © 1995 The Open University.

All rights reserved. No part of this publication may be reproduced, stored in a retrieval system or transmitted, in any form or by any means, without written permission from the publisher or a licence from the Copyright Licensing Agency Limited. Details of such licences (for reprographic reproduction) may be obtained from the Copyright Licensing Agency Ltd of 90 Tottenham Court Road, London, W1P 9HE.

Designed by the Graphic Design Group of the Open University.

Typeset in the United Kingdom by the Open University.

Printed in the United Kingdom by Hobbs the Printers Ltd, Totton, Hampshire, SO40 3WX.

ISBN 0 7492 2221 2

This text forms part of an Open University third-level course. If you would like a copy of *Studying with the Open University*, please write to the Central Enquiry Service, PO Box 200, The Open University, Walton Hall, Milton Keynes, MK7 6YZ. If you have not enrolled on the course and would like to buy this or other Open University material, please write to Open University Educational Enterprises Ltd, 12 Cofferidge Close, Stony Stratford, MK11 1BY, United Kingdom.

MT365Graphs2i1.3

Contents

Introduction		4
1 Tree structures		4
	1.1 Mathematical properties of trees	4
	1.2 Examples of trees	6
	1.3 Modelling with trees	13
2 Counting trees		15
	2.1 Counting labelled trees	16
	2.2 Counting binary trees	21
	2.3 Counting chemical trees	23
	2.4 Postscript	25
3 Greedy algorithms		26
	3.1 Spanning trees	26
	3.2 Minimum connector problem	27
	3.3 Maximum connector problem	34
	3.4 Travelling salesman problem	36
	3.5 Computer activities	38
4 Multi-terminal flows		39
	4.1 Cut trees	39
	4.2 Algorithm of Gomory and Hu	45
5 Gas pipeline networks		46
Appendix: Trees, telephones and tiles		46
Further reading		51
Exercises		52
Solutions to the exercises		55
Solutions to the problems		64
Index		75

Introduction

In Block 1 we introduced the main areas of this course in general terms. In particular, in *Graphs 1*, our discussion of graphs was designed to emphasize those features which embody the essence of the concept of a graph, and our examples were chosen to illustrate the wide range of situations which can be modelled by graphs.

In this unit we narrow our horizons and focus our attention on one particularly important and useful type of graph — a *tree*. Although trees are relatively simple structures, they form the basis of many of the practical techniques used to model and to design large-scale systems.

We start, in Section 1, *Tree structures*, by introducing some useful concepts and results that will be needed later. We then examine a number of situations in which tree structures arise. In particular, we describe some applications in which a hierarchical, or branching, representation is the most natural one, and discuss some advantages and disadvantages of using trees to model various situations.

In Section 2, *Counting trees*, we discuss some specific tree-counting problems; one of these, the counting of chemical trees, was important in the historical development of graph theory.

Section 3, *Greedy algorithms*, is devoted to a discussion of algorithms used for constructing a particular type of tree called a *spanning tree*. Such algorithms are useful for a number of optimization problems, including the *minimum connector problem* you met in the *Introduction* unit.

In Section 4, *Multi-terminal flows*, we discuss the problem of efficiently determining maximum flows between a number of terminals in an undirected network, and we introduce a special type of tree called a *cut tree*.

Finally, in the television section, Section 5, *Gas pipeline networks*, we show how tree networks are used in the construction of cost-effective gas pipeline distribution systems.

1 Tree structures

The concept of a *tree* is one of the most important and commonly used ideas in graph theory, especially in the applications of the subject. It arose in connection with the work of Gustav Kirchhoff on electrical networks in the 1840s, and later with Arthur Cayley's work on the enumeration of molecules in the 1870s. More recently, trees have proved to be of value in such areas as computer science, decision making, linguistics, and the design of gas pipeline systems.

In this section we first investigate the mathematical properties of trees and then look at some applications.

1.1 Mathematical properties of trees

We start by recalling the definition of a tree.

> **Definition**
>
> A **tree** is a connected graph that has no cycles.

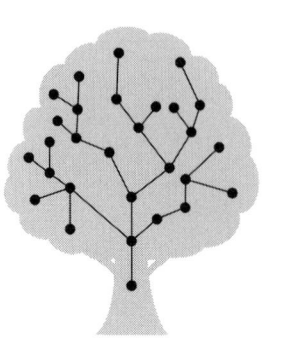

For example, the following diagram depicts all the unlabelled trees with at most five vertices.

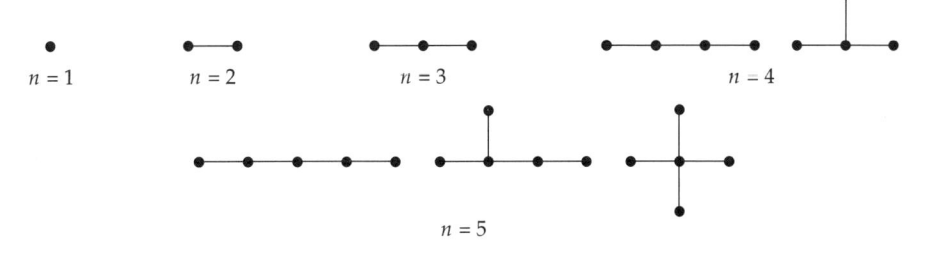

•   •—•   •—•—•   •—•—•—•   •—•—•
$n = 1$ $n = 2$ $n = 3$ $n = 4$

$n = 5$

Problem 1.1

Draw the six unlabelled trees with six vertices.

Note that each unlabelled tree with six vertices can be obtained from an unlabelled tree with five vertices by adding an edge joining a new vertex to an existing vertex. This is a general procedure for increasing the size of a tree, since it creates no cycles and can be carried out systematically by adjoining the new edge to each vertex in turn. For example, consider the following tree with six vertices.

By adjoining a new edge to each vertex in turn, we obtain the following trees with seven vertices.

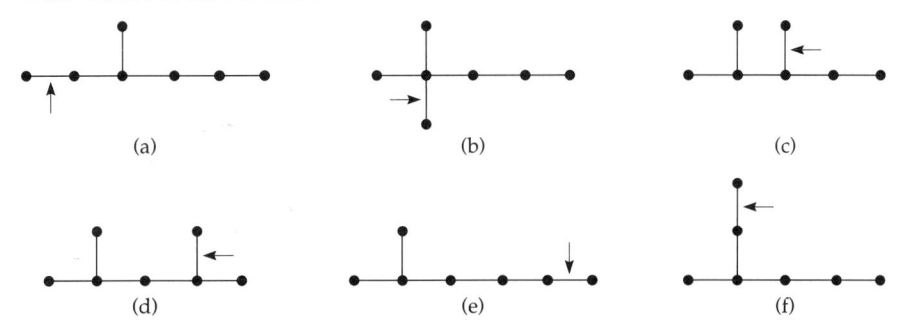

(a)   (b)   (c)

(d)   (e)   (f)

We can omit tree (f) from this list, since it is isomorphic to tree (a), and so we obtain five trees with seven vertices from our original tree with six vertices. Note that the difficulty of producing trees in this way is in recognizing duplicates, but at least we can be sure that each tree with seven vertices can be thus obtained from some tree with six vertices.

Problem 1.2

By adding a new edge in all possible ways to each unlabelled tree with six vertices, draw the eleven unlabelled trees with seven vertices.

Starting with the tree with just one vertex, we can build up any tree we wish by successively adding a new edge and a new vertex. At each stage, the number of vertices exceeds the number of edges by 1, so that

every tree with n vertices has exactly $n - 1$ edges.

At no stage is a cycle created, since each added edge joins an old vertex to a new vertex.

Note that, at each stage, the tree remains connected, and so any two vertices must be connected by at least one path. However, they cannot be connected by more than one path, since any two such paths would contain a cycle (and possibly other edges as well). We can therefore deduce that

any two vertices in a tree are connected by exactly one path.

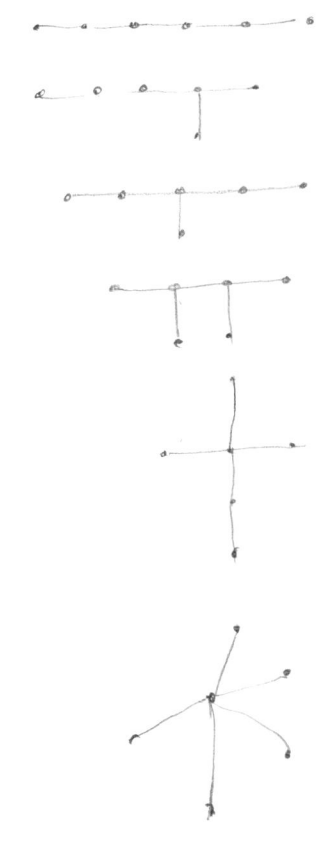

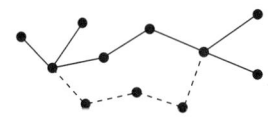

In particular, any two *adjacent* vertices are connected by exactly one path — the edge joining them. If this edge is removed, then there is no path between the two vertices. It follows that

> the removal of any edge of a tree disconnects the tree.

Moreover,

> joining any two vertices of a tree by an edge creates a cycle.

This is because any two vertices v and w are connected by a path, and the addition of the edge vw produces a cycle — the cycle consisting of the path and the added edge vw.

Several of the above properties can be used as alternative definitions of a tree. In the following theorem, we state six alternative definitions. They are all *equivalent*: any one of them can be taken as the definition of a tree, and the other five can then be deduced.

Theorem 1.1: equivalent definitions of a tree

Let T be a graph with n vertices. Then the following statements are equivalent.

- T is connected and has no cycles.
- T has $n-1$ edges and has no cycles.
- T is connected and has $n-1$ edges.
- T is connected and the removal of any edge disconnects T.
- Any two vertices of T are connected by exactly one path.
- T contains no cycles, but the addition of any new edge creates a cycle.

Problem 1.3

(a) Show that the removal of an edge cannot disconnect a tree into more than two components. Use a proof by contradiction.

(b) Show that the addition of a new edge to a tree cannot create more than one cycle. Use a proof by contradiction.

Problem 1.4

(a) Give an example of a tree with seven vertices and

  (1) exactly two vertices of degree 1;

  (2) exactly four vertices of degree 1;

  (3) exactly six vertices of degree 1.

(b) Use the handshaking lemma to prove that every tree with n vertices, where $n \geq 2$, has at least two vertices of degree 1. Use a proof by contradiction.

The handshaking lemma was stated in *Graphs 1*, Section 1.

1.2 Examples of trees

In Section 1.1 we defined the concept of a tree. Such trees can often be used to model situations involving various physical or conceptual tree-like structures. These structures are also commonly referred to as 'trees'. In the following examples, we classify such 'trees' in terms of the type of application in which they occur. In particular, we distinguish between structures which are realized physically, and those which are realized conceptually. In the following discussion, our treatment is mainly pictorial and intuitive.

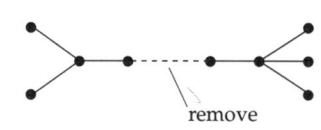

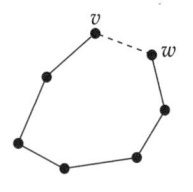

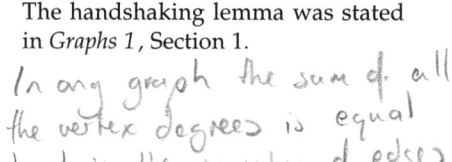

Physical trees

Many trees have a *physical* structure which may be either natural or artificial and either static or time-dependent. Two examples of natural trees are the biological variety with trunk, branches and leaves, and the drainage system of tributaries forming a river basin. Less obvious examples of tree structures are provided by the tracks of elementary particles in a bubble chamber, and by the chemical structure of certain organic molecules.

A system of tributaries is often tree-like in appearance, although it ceases to have a pure tree structure if the tributaries recombine.

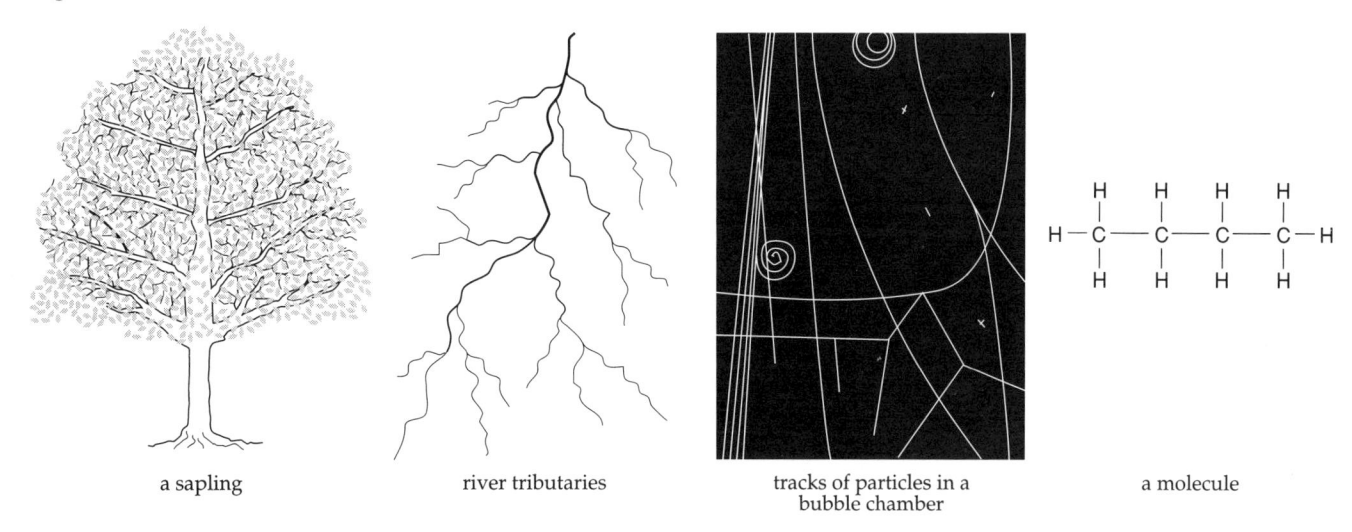

a sapling

river tributaries

tracks of particles in a bubble chamber

a molecule

An example of the artificial variety of tree is an oil or gas pipeline distribution system, such as an undersea pipeline network of the type considered in Section 5; since the cost of constructing such a network may be very large, a tree structure with no unnecessary edges may be the most economical form for the network.

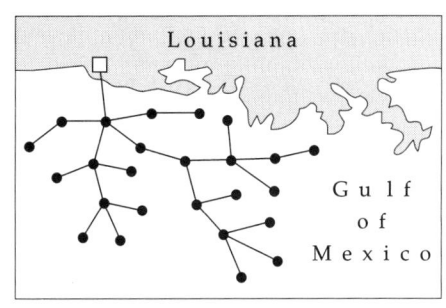

a pipeline distribution system

Conceptual trees

Many trees do not have a well-defined physical structure, but are *conceptual*. Probably the most familiar type of conceptual tree is a *family tree*, depicting ancestors and descendants. The following diagram illustrates part of the family tree of Saxon kings in the ninth century in England.

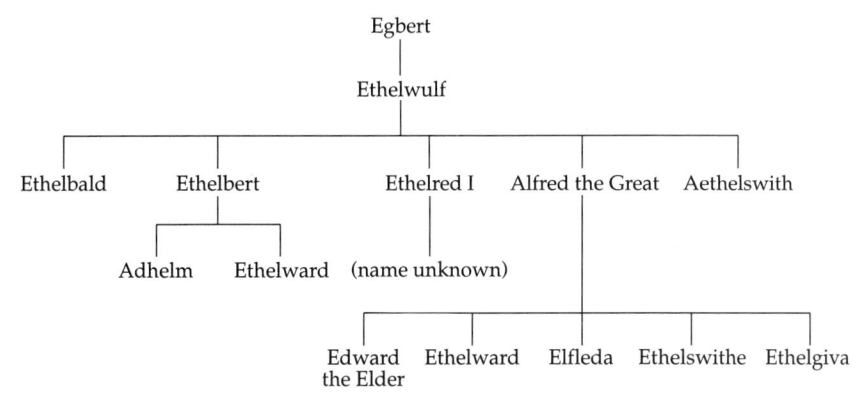

Note that a family tree has a pure tree form only if the branches never recombine; if interbreeding takes place, then there are cycles in the graph and we obtain a structure called a *semi-lattice*. These are discussed in Section 1.3.

There are many other instances of conceptual trees. Examples of these are trees used to represent administrative hierarchies; the following tree is one representation of the administrative structure of the Operations area of the Open University.

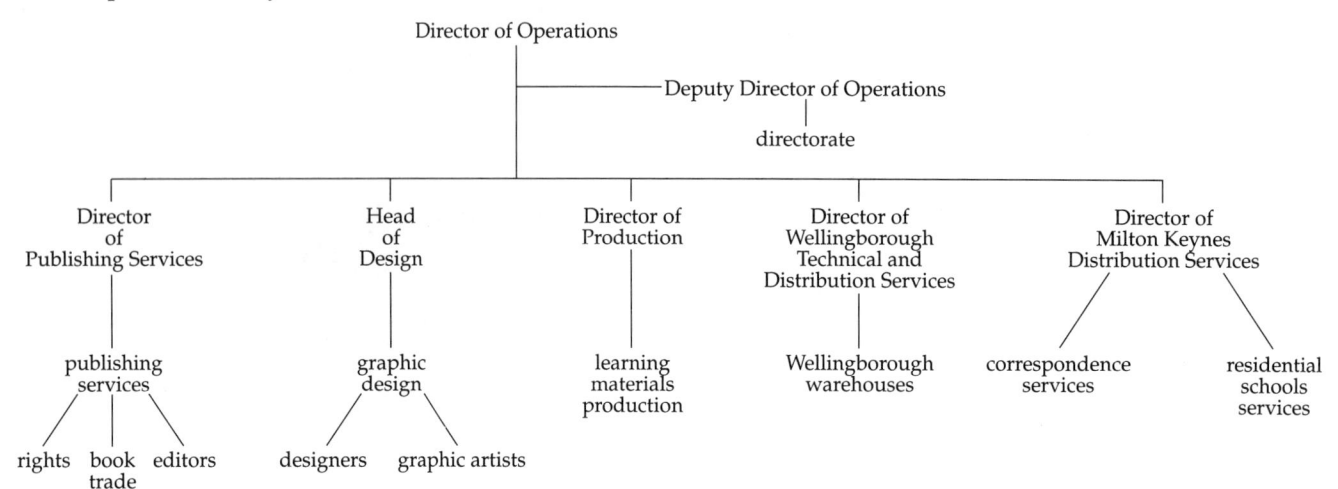

```
                          Director of Operations
                                  |
                                  |────Deputy Director of Operations
                                  |          |
                                  |       directorate
    ──────────┬──────────┬────────┼──────────────┬──────────────────┐
  Director   Head    Director of   Director of         Director of
     of       of      Production   Wellingborough      Milton Keynes
  Publishing Design                Technical and       Distribution Services
  Services                         Distribution Services
     |         |          |              |              /        \
  publishing graphic   learning     Wellingborough  correspondence  residential
  services   design    materials    warehouses      services        schools
   /|\        /\       production                                   services
rights book editors  designers graphic artists
      trade
```

Another hierarchical tree structure is described in the rhyme from Augustus De Morgan's *A Budget of Paradoxes* (adapted from an earlier poem of Jonathan Swift).

Great fleas have little fleas
    upon their backs to bite 'em,
And little fleas have lesser fleas,
    and so *ad infinitum*.

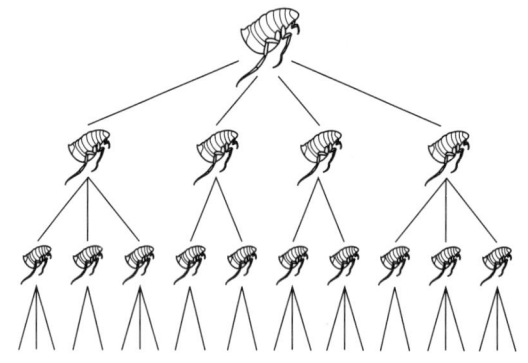

Problem 1.5

Describe the tree of objects that I met in the following rhyme.
    As I was going to St. Ives,
    I met a man with seven wives,
    Each wife had seven sacks,
    Each sack had seven cats,
    Each cat had seven kits, ...

Rooted trees

Among the examples of tree structures, one particular type of tree occurs repeatedly. This is the hierarchical structure in which one vertex is singled out as the starting point, and the branches fan out from this vertex. We call such trees **rooted trees**, and refer to the starting vertex as the **root**. For example, the above tree representing the administrative structure of the Operations area of the Open University is a rooted tree, with the Director of Operations as the root.

A rooted tree is often drawn as follows, with the root indicated by a small square at the top, and the various branches descending from it. When a path from the top reaches a vertex, it may split into several new branches. A rooted tree in which there are at most *two* branches at any vertex is a **binary tree**.

a rooted (branching) tree *a binary tree*

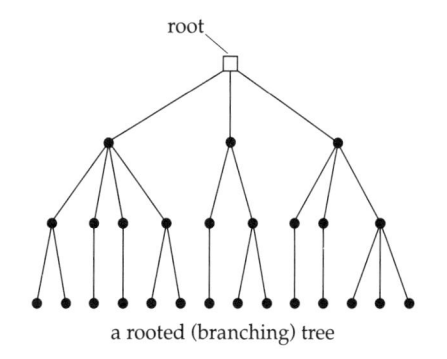

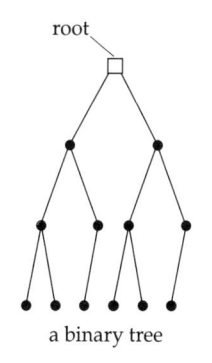

Note that, although a top-to-bottom direction is often implied, we usually draw a rooted tree as a graph with undirected edges, rather than as a digraph with arcs directed downwards.

Because of the branching that takes place at the vertices, such trees are often called **branching trees**. We have already seen instances of branching trees — the family tree and the Open University tree, for example. There are many further examples, as we now show.

Outcomes of experiments

If we toss a coin or throw a die several times, then the possible outcomes can be represented by a branching tree. In the case of tossing a coin, each possible outcome has two edges leading from it, since the next toss may be a head (H) or a tail (T), and we obtain a binary tree. For example, if we toss a coin three times, then there are eight possible outcomes, and we obtain the following branching tree.

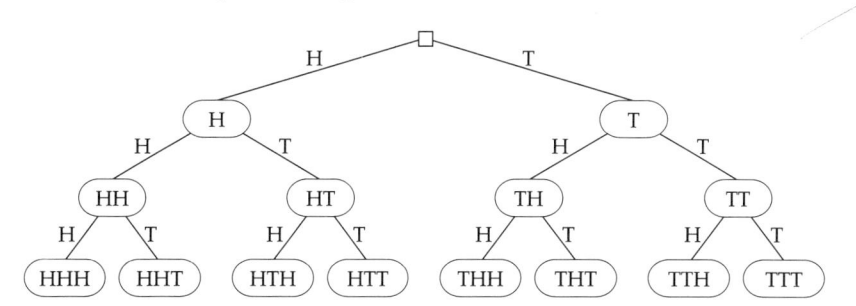

Problem 1.6

Draw the branching tree representing the outcomes of two throws of a six-sided die.

Games of strategy

Branching trees arise in the analysis of games, particularly games of strategy such as chess and noughts-and-crosses, or for strategic manoeuvres such as those arising in military situations. In such tree representations, a path from the top corresponds to a sequence of moves from each position to the next. The following diagram illustrates the branching tree representing the first three moves in a game of noughts-and-crosses.

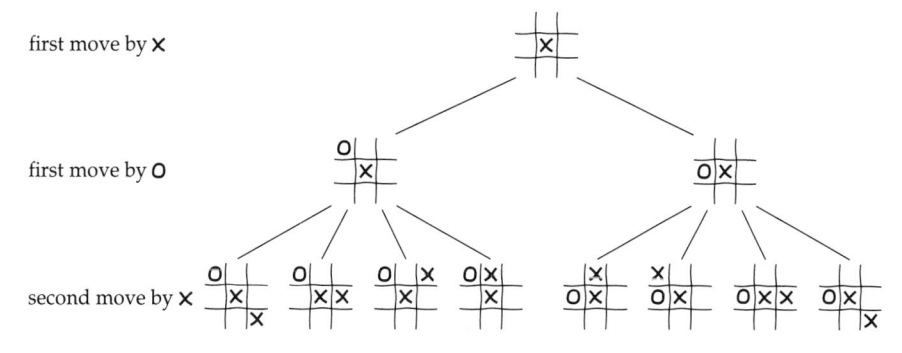

first move by **X**

first move by **O**

second move by **X**

Equivalent moves such as:

and

are considered to be the same.

9

If, however, we are concerned with the positions of the game, rather than with the sequence of moves, then certain paths may 'close up' into cycles, and we no longer obtain a tree.

The type of graph we obtain is a semi-lattice.

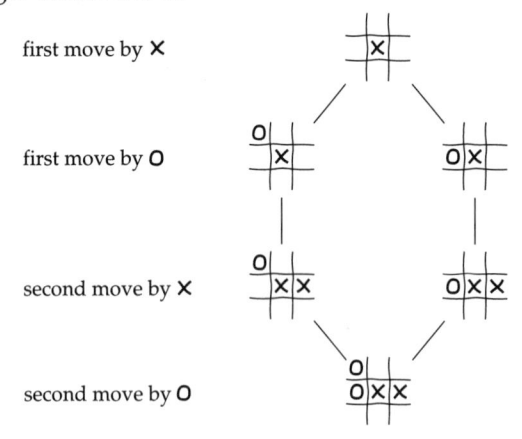

Grammatical trees

Branching trees occur in the parsing of a sentence in a natural language, such as English. The tree represents the interrelationships between the words and phrases of the sentence, and hence the underlying syntactic structure. Such a branching tree is obtained by splitting the sentence into noun phrases and verb phrases, then splitting these phrases into nouns, verbs, adjectives, and so on. For example, the structure of the sentence *Good students read books* can be represented by the following tree.

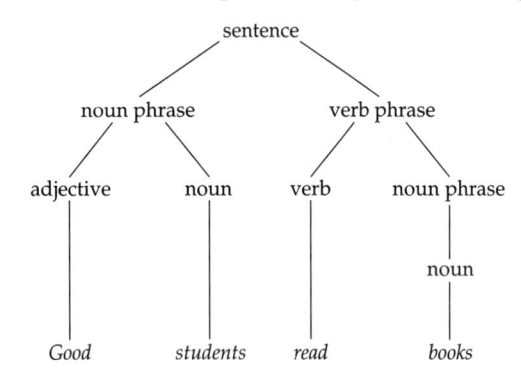

If a sentence is ambiguous, we can use branching trees to distinguish between the different sentence constructions. For example, the newspaper headline *Council rents rocket* can be interpreted in two ways, as illustrated by the following trees.

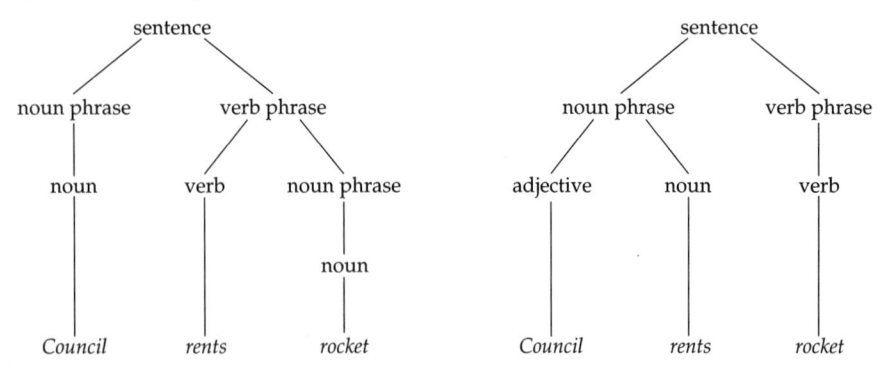

Problem 1.7

Use a branching tree to parse the sentence *Robin wears red socks*.

Computer science

Rooted tree structures arise in computer science, where they are used to model and describe branching procedures in programming languages (the languages used to write algorithms to be interpreted by computers). In particular, they are used to store data in a computer's memory in many different ways. For example, consider the list of seven numbers 7, 5, 4, 2, 1, 6, 8. The following trees represent ways of storing this list in the memory — as a *stack* and as a *binary tree*. Each representation has its advantages, depending on how the data is to be manipulated, but in both representations it is important to distinguish where the data starts, so the trees are *rooted* trees.

A discussion of data storage and binary search trees is given in *Graphs 4, Graphs and computing*.

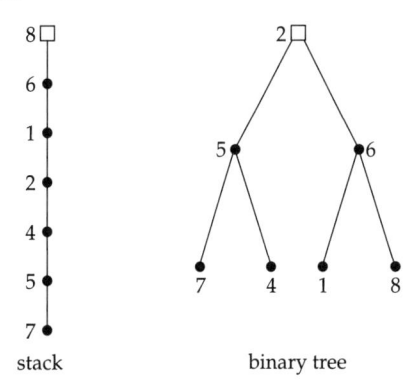

We obtain the tree by writing the numbers in a string 7542168, 'promoting' every second number (5,2,6) and then 'promoting' the second number (2).

stack   binary tree

Sorting trees

Sorting trees are branching trees that arise when we wish to make a succession of choices, each dependent on the previous one. For example, the Dewey decimal classification system, developed by Melvil Dewey in the nineteenth century, has often been used for cataloguing library books.

There is an initial classification of subjects into ten areas:

0–099	General works	500–599	Pure sciences
100–199	Philosophy	600–699	Applied sciences
200–299	Religion	700–799	Fine arts
300–399	Social sciences	800–899	Literature
400–499	Languages	900–999	History

With the proliferation of new subjects that cannot be classified conveniently or unambiguously in this way, the Dewey system is increasingly being augmented by computer systems that interrelate the various areas more adequately.

Each of these areas is classified into ten more specialized areas — for example, in the pure sciences the range 510–519 is allocated to mathematics, which is then further subdivided as:

510	General mathematical works	515	Analysis
511	Mathematical foundations	516	Geometry
512	Algebra	517	(unassigned)
513	Arithmetic	518	(unassigned)
514	Topology	519	Probability

Further classification necessitates the introduction of decimal fractions. For example, graph theory is classified as 511.5 and combinatorial analysis as 511.6. To represent this classification process, we can use a sorting tree, part of which is:

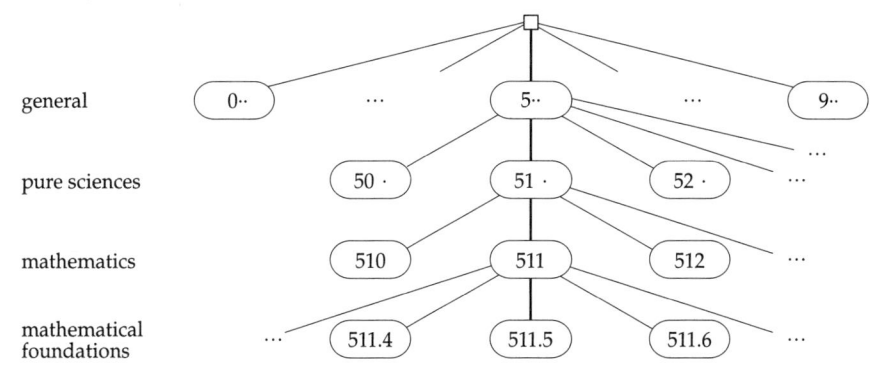

Problem 1.8

Explain how a sorting tree can be used to represent the sorting of mail according to postcode.

Equivalent forms

We conclude this subsection by showing how rooted trees can be represented in several different ways. Because such trees are important and widespread, we need to be able to recognize these different forms.

The following diagrams illustrate four equivalent ways of representing the same rooted tree.

| a rooted tree | subsets of a set | nested parentheses | sections of a report |
| (a) | (b) | (c) | (d) |

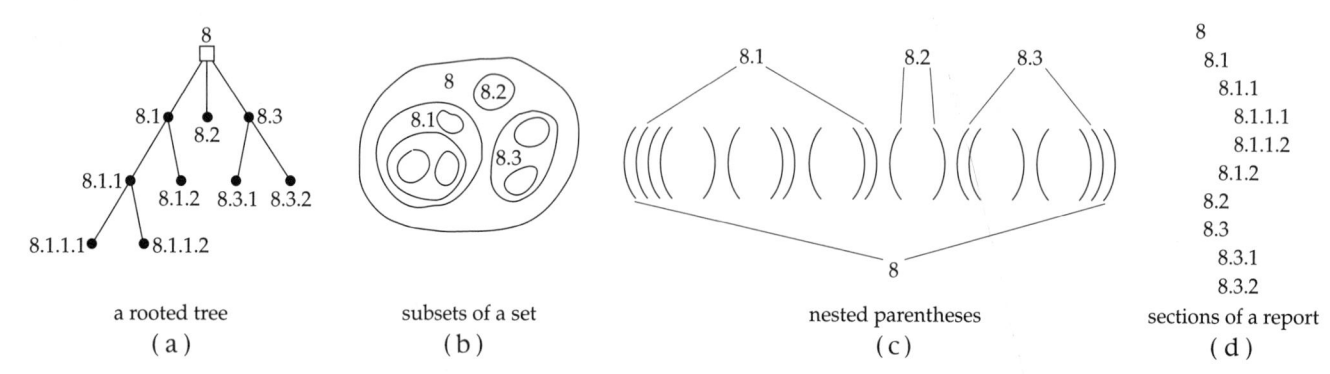

Diagram (a) has the conventional appearance of a *rooted tree*.

Diagram (b) is a system of *subsets of a set* representing, say, the organization of subsystems within a complex machine system; this has the same tree structure as the first diagram, but in this case the different levels are defined by the depth of nesting.

Diagram (c) is a system of *nested parentheses* as used in English text, mathematical equations or the computer language LISP; again, the level is defined by the depth of nesting.

Diagram (d) is provided by the organization of a report such as a government report or a legal contract; these are often arranged in *nested sections* (subsections, paragraphs, etc.), and the level of each section is indicated by indentation and by the length of the decimal number in the heading. Books are frequently organized in this tree-like way (volumes, chapters, sections, paragraphs), as are dictionaries (the lexicographical ordering of words).

Problem 1.9

Write down the corresponding subsets of a set and nested parentheses for the following rooted tree.

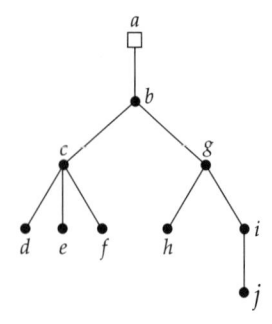

The advantage of tree structures is the ease with which they can be altered or updated. This is particularly important in computer applications, where we can insert or delete branches (such as subroutines) without having to change the whole system. On the other hand, a major drawback of tree structures is that they can be vulnerable to faults or damage. The removal of a single vertex or the breaking of a single edge is sufficient to disconnect or destroy the system, which can be disastrous for efficient operation. A striking example of this vulnerability was given by the collapse of the Inca civilization which virtually disintegrated overnight when the Spanish conquistador Pizarro captured the chief Inca, Emperor Atahuallpa, in 1532. The latter occupied the top position in a rigid hierarchical social pyramid, and his removal destroyed the root of the tree, thereby breaking the chain of command.

1.3 Modelling with trees

Many situations can be modelled by trees; in particular, hierarchical situations can frequently be modelled by rooted trees. However, for many practical purposes, the concept of a tree is too simplistic to be used as a model, and a less obvious and more complicated structure is needed. For example, we mentioned above that several structures are represented more adequately by a *semi-lattice*.

For example, consider the following hierarchical tree, representing the lines of responsibility in a given company.

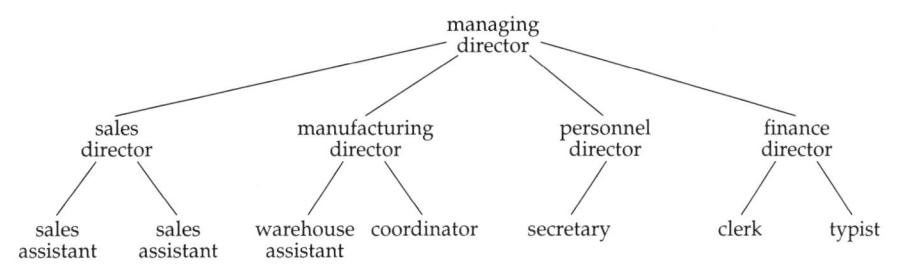

Such a representation in tree form is both simple and convenient, and may be perfectly satisfactory for some purposes. However, there may be instances where a sales assistant has a problem that cannot be dealt with by his/her superior, the sales director, who needs to refer it to the finance director. Equally, the four directors on the middle level may need to become involved with each other's activities, and it may be necessary to draw some horizontal edges at this level. Thus the simple tree structure breaks down, as extra linking edges are introduced. In such circumstances, it is essential that we understand exactly what the presence of an edge is meant to represent, and that we do not try to impose a tree structure where it is inappropriate.

As another example, we may wish to draw a tree structure representing the classification of animals. It has been found useful in evolutionary biology to separate the set of all animals into those that lay eggs and those that do not. Similarly, animals that suckle their young form another class, and we would expect to have a tree that looks something like the one on the left:

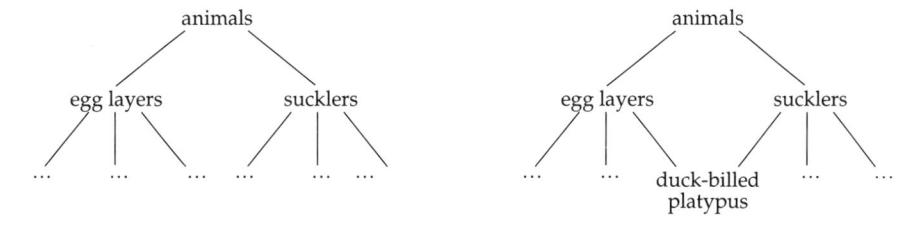

Unfortunately, in drawing this tree, we have forgotten that the duck-billed platypus is both an egg layer and a suckler, and consequently our pure tree structure is spoiled by the presence of a cycle, as shown above on the right. Exceptions such as the duck-billed platypus cannot be 'wished away': it is the mathematical representation that must be changed.

In this case, we obtain a semi-lattice.

A city is not a tree

Such considerations were expressed very vividly in 1966 by Christopher Alexander, who wrote an influential article entitled *A city is not a tree*. In this article, he described the unsatisfactory nature of what he called *artificial cities*; examples of these are the new towns built in many countries since the War, such as Chandigarh in India, Levittown in Pennsylvania, and some of the British new towns. Contrasting these with *natural cities*, such as Liverpool, Kyoto and Manhattan, he remarked:

> It is more and more widely recognised today that there is some essential ingredient missing from artificial cities. When compared with ancient cities that have acquired the patina of life, our modern attempts to create cities artificially are, from a human point of view, entirely unsuccessful.

To support his views, Alexander analysed the two types of city from a structural point of view, and concluded that modern cities often have a tree structure that leads to a rather sterile environment. For example, in Columbia, Maryland, neighbourhoods are arranged in clusters of five to form 'villages'. Transportation links the villages into a new town; the structure can be represented by a tree.

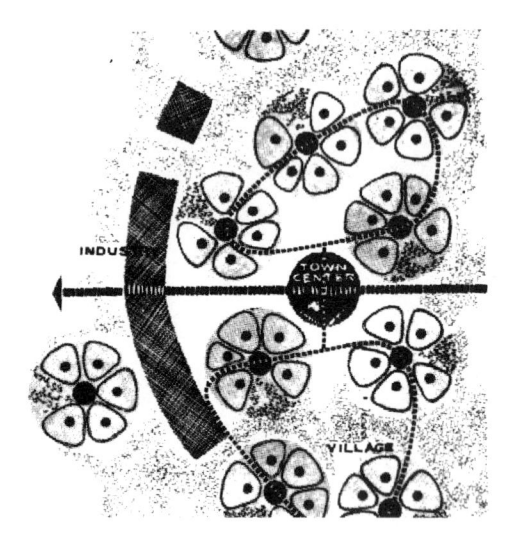

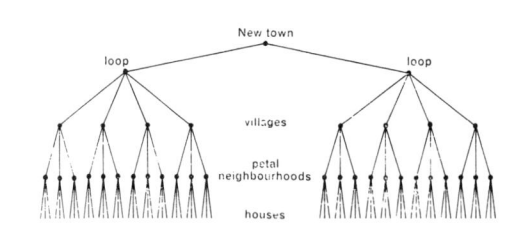

Columbia, Maryland

Another example is Chandigarh, planned by Le Corbusier in 1951. The whole city is served by a commercial centre in the middle, linked to the administrative centre. Two subsidiary elongated commercial cores are strung out along the major arterial roads, running from north to south. Subsidiary to these are further administrative community and commercial centres, one for each of the city's twenty sectors.

By way of contrast, natural cities tend to have everything mixed up together, and provide a richer environment: the grocer's shop is just round the corner, the theatre is a short walk from home, there is a nearby restaurant, and so on. The graph that represents the structure of such a city is more complicated than a tree.

The graphs arising from natural cities in this way are called **semi-lattices**, and have the property that, if two vertices correspond to areas of a map that overlap, then the overlap area should also be regarded as an area of interest in its own right and be represented by a corresponding vertex.

Alexander discussed a redevelopment plan for the city of Middlesborough, in which the city was partitioned into twenty-nine neighbourhoods. The idea was that each neighbourhood should have its own services, such as elementary schools, secondary schools, youth clubs, adult clubs and post offices. The expectation was that people would use the facilities in their own neighbourhood, as in the tree structure shown below on the left. In fact, this was not the case. Although each facility had its own catchment area, they overlapped, giving rise to the more complicated graph — a semi-lattice — shown below on the right.

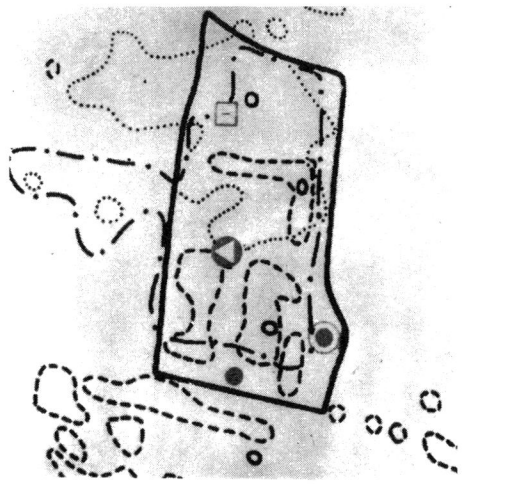

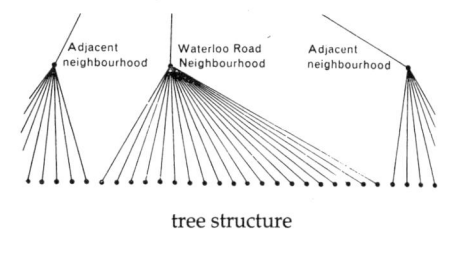

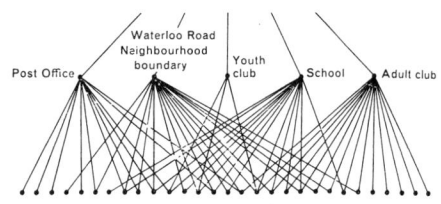

development plan for Middlesborough

tree structure

graph structure

It follows from the above discussion that, while a rooted tree is a simple model that can usefully be employed to represent a number of practical situations, it is often too simple to be of any direct use. In such cases, we can sometimes model the situation with a semi-lattice in which some of the downward paths combine to give cycles.

After studying this section, you should be able to:

- state several properties of a tree, and give several equivalent definitions of a tree;
- distinguish between physical and conceptual tree structures, and give examples of each type;
- appreciate the uses of rooted trees in a number of different areas;
- appreciate the limitations of modelling with trees, and describe situations that are modelled by semi-lattices, rather than by trees.

---

2 Counting trees

Much of the interest and importance of trees arises from the fact that, in many ways, a tree is the simplest non-trivial type of graph. Consequently, when investigating a problem in graph theory, it is sometimes convenient to start by investigating the corresponding problem for trees.

In this section we turn our attention to the enumeration problem of counting trees with particular properties. Determining the number of graphs with a given property is often very difficult, but the corresponding problems for trees are usually much easier. Furthermore, we can sometimes use the solutions to many standard tree-counting problems to deduce the solutions to other counting problems.

Two typical enumeration problems are given below.

> How many irrigation canal systems are there linking eight locations with seven canals?

> How many molecules are there with the formula C_8H_{18}?

We can reduce many such problems to that of determining the number of trees with a particular property. For example, the first problem amounts to counting labelled trees with eight vertices; the second reduces to that of determining the number of unlabelled trees with eight vertices, each with degree not exceeding 4.

2.1 Counting labelled trees

As you saw in *Graphs 1*, counting problems for labelled graphs are usually much easier to solve than their analogues for unlabelled graphs; in fact, there are certain types of graph for which the former problem has been solved while the latter problem remains unsolved. However, the problems of counting labelled and unlabelled trees have both been solved, although the former problem is much easier to solve than the latter.

The following table lists the numbers of unlabelled and labelled trees with n vertices, where $n \leq 9$. The labelled case illustrates the *combinatorial explosion*, mentioned in the *Introduction* unit.

n	1	2	3	4	5	6	7	8	9
unlabelled trees	1	1	1	2	3	6	11	23	47
labelled trees	1	1	3	16	125	1296	16 807	262 144	4 782 969

The labelled trees with at most three vertices are as follows.

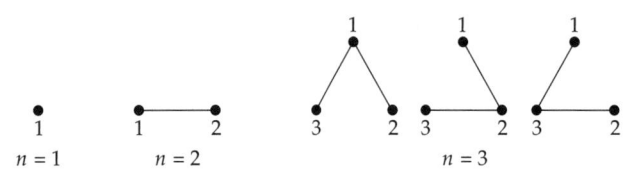

Problem 2.1

Draw the sixteen labelled trees with four vertices.

Hint Draw the two unlabelled trees with four vertices as $K_{1,3}$ and P_4, and label them in all possible ways.

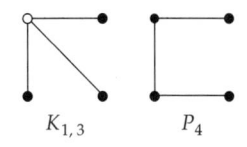

Problem 2.2

Using the above table, try to guess a simple formula for the number of labelled trees with n vertices.

The fact that there are exactly n^{n-2} labelled trees with n vertices is known as *Cayley's theorem*. To prove this result, we construct a one–one correspondence between labelled trees with n vertices and sequences of $n - 2$ numbers, called **Prüfer sequences**.

This construction is due to H. Prüfer.

The construction of the one–one correspondence between labelled trees and Prüfer sequences is in two parts.

In the first part, given a labelled tree with n vertices, we construct a Prüfer sequence

$(\mathbf{a}_1, \mathbf{a}_2, \mathbf{a}_3, \ldots, \mathbf{a}_{n-2})$,

We use **bold type** for the terms of a Prüfer sequence.

where each $\mathbf{a}_i$ is one of the integers $1, 2, 3, \ldots, n$ (allowing repetition); for example, if $n = 7$, then two of the possible sequences are $(\mathbf{1}, \mathbf{2}, \mathbf{3}, \mathbf{4}, \mathbf{5})$ and $(\mathbf{7}, \mathbf{7}, \mathbf{7}, \mathbf{1}, \mathbf{2})$. This type of construction is shown in Construction A and Example 2.1A.

Construction A: to construct a Prüfer sequence from a given labelled tree

STEP 1 Find the vertices of degree 1 and choose the one with the smallest label.

STEP 2 Look at the vertex adjacent to the one just chosen, and place its label in the first available position in the Prüfer sequence.

STEP 3 Remove the vertex chosen in Step 1 and its incident edge, leaving a smaller tree.

Repeat Steps 1–3 for the remaining tree, continuing until there are only two vertices left, then STOP: the required Prüfer sequence has been constructed.

Example 2.1A

Consider the following labelled tree.

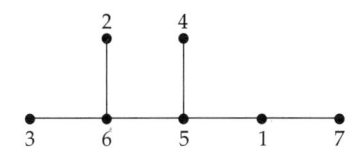

Since this labelled tree has 7 vertices, the corresponding Prüfer sequence has 5 numbers.

First term   *Prüfer sequence*

STEP 1 The vertices of degree 1 are vertices 3, 2, 4 and 7; the one with the smallest label is vertex 2.

STEP 2 The vertex adjacent to vertex 2 is vertex 6, so the first term in the sequence is **6**.   (**6**, ·, ·, ·, ·)

STEP 3 Removal of vertex 2 and edge 26 leaves the following tree.

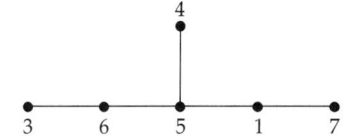

Second term

STEP 1 The vertices of degree 1 are vertices 3, 4 and 7; the one with the smallest label is vertex 3.

STEP 2 The vertex adjacent to vertex 3 is vertex 6, so the second term in the sequence is **6**.   (**6, 6**, ·, ·, ·)

STEP 3 Removal of vertex 3 and edge 36 leaves the following tree.

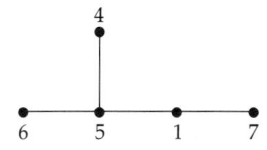

Third term

STEP 1 The vertices of degree 1 are vertices 4, 6 and 7; the one with the smallest label is vertex 4.

STEP 2 The vertex adjacent to vertex 4 is vertex 5, so the third term in the sequence is **5**.   (**6, 6, 5**, ·, ·)

STEP 3 Removal of vertex 4 and edge 45 leaves the following tree.

Fourth term

STEP 1   The vertices of degree 1 are vertices 6 and 7; the one with the smaller label is vertex 6.

STEP 2   The vertex adjacent to vertex 6 is vertex 5, so the fourth term in the sequence is **5**.   (6, 6, 5, 5, ·)

STEP 3   Removal of vertex 6 and edge 65 leaves the following tree.

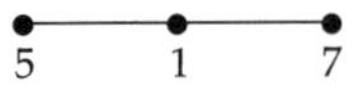

•———•———•
5   1   7

Fifth term

STEP 1   The vertices of degree 1 are vertices 5 and 7; the one with the smaller label is vertex 5.

STEP 2   The vertex adjacent to vertex 5 is vertex 1, so the next term in   (6, 6, 5, 5, 1)
the sequence is **1**.

STEP 3   Removal of vertex 5 and edge 51 leaves a tree with only two vertices.

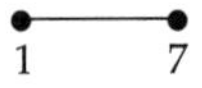

•———•
1   7

STOP.

We thus obtain the Prüfer sequence (**6, 6, 5, 5, 1**). ■

Problem 2.3

Find the Prüfer sequence corresponding to each of the following labelled trees.

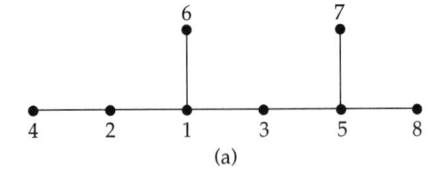

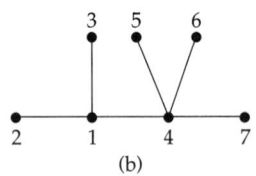

(a) tree with vertices 4—2—1—3—5—8, with 6 attached above 1 and 7 attached above 3.

(b) tree with vertices 2—1—4—7, with 3, 5, 6 attached above 4.

In the second part of the construction of the one–one correspondence between labelled trees and Prüfer sequences, we give the converse of Construction A.

In this second part, given a sequence ($a_1, a_2, a_3, \ldots, a_{n-2}$), where each a_i is one of the integers 1, 2, 3, ..., n (allowing repetition), we construct a labelled tree with n vertices. This type of construction is shown in Construction B and Example 2.1B.

Construction B: to construct a labelled tree from a given Prüfer sequence

STEP 1   Draw the n vertices, labelling them from 1 to n, and make a list of the integers from 1 to n.

STEP 2   Find the smallest number which is in the list but not in the Prüfer sequence, and also find the first number in the sequence; then add an edge joining the vertices with these labels.

STEP 3   Remove the first number found in Step 2 from the list and the second number found in Step 2 from the sequence, leaving a smaller list and a smaller sequence.

Repeat Steps 2 and 3 for the remaining list and sequence, continuing until there are only two terms left in the list. Then join the vertices with these labels and STOP: the required labelled tree has been constructed.

Example 2.1B

Consider the Prüfer sequence (**6, 6, 5, 5, 1**); since this sequence has five numbers, the corresponding labelled tree has 7 vertices.

labelled tree

No edges

STEP 1 Draw the 7 vertices, labelling them 1 to 7, and list the integers from 1 to 7.

The list is (1, 2, 3, 4, 5, 6, 7) and the sequence is (**6, 6, 5, 5, 1**).

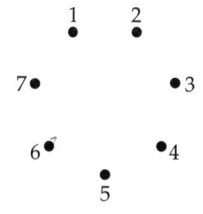

First edge

STEP 2 The smallest number in the list but not in the sequence is 2, and the first number in the sequence is **6**, so we add an edge joining vertices 2 and 6.

STEP 3 We remove the number 2 from the list and the number **6** from the sequence.

This leaves the list (1, 3, 4, 5, 6, 7) and the sequence (**6, 5, 5, 1**).

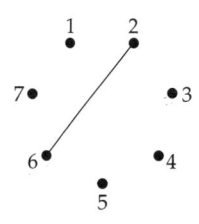

Second edge

STEP 2 The smallest number in the new list that is not in the new sequence is 3, and the first number in the new sequence is **6**, so we add an edge joining vertices 3 and 6.

STEP 3 We remove the number 3 from the list and the number **6** from the sequence.

This leaves the list (1, 4, 5, 6, 7) and the sequence (**5, 5, 1**).

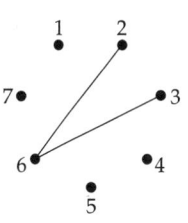

Third edge

STEP 2 The smallest number in the new list that is not in the new sequence is 4, and the first number in the new sequence is **5**, so we add an edge joining vertices 4 and 5.

STEP 3 We remove the number 4 from the list and the number **5** from the sequence.

This leaves the list (1, 5, 6, 7) and the sequence (**5, 1**).

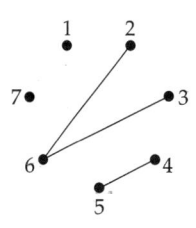

Fourth edge

STEP 2 The smallest number in the new list that is not in the new sequence is 6, and the first number in the new sequence is **5**, so we add an edge joining vertices 6 and 5.

STEP 3 We remove the number 6 from the list and the number **5** from the sequence.

This leaves the list (1, 5, 7) and the sequence (**1**).

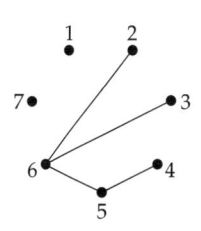

Fifth edge

STEP 2 The smallest number in the new list that is not in the new sequence is 5, and the number in the new sequence is **1**, so we add an edge joining vertices 5 and 1.

STEP 3 We remove the number 5 from the list and the number **1** from the sequence.

This leaves the list (1, 7) and an empty sequence.

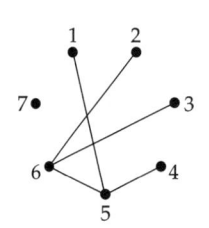

Sixth edge

>The labels 1 and 7 are the only two terms left in the list, so we join vertices 1 and 7.

STOP.

We thus obtain the required labelled tree.

Redrawing the tree without edges crossing, we obtain the following.

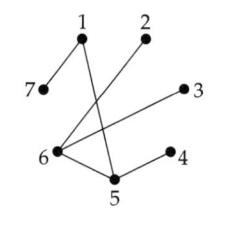

Problem 2.4

Find the labelled tree corresponding to each of the following Prüfer sequences.

(a) **(2, 1, 1, 3, 5, 5)**   (b) **(1, 1, 4, 4, 4)**

What do you notice about your results?

Note that, both in the above examples and in Problems 2.3 and 2.4, the Prüfer sequence arising from a particular labelled tree in Construction A gives rise to the same labelled tree in Construction B; for example, the Prüfer sequence **(6, 6, 5, 5, 1)** from the labelled tree in Example 2.1A gives rise to the same labelled tree in Example 2.1B.

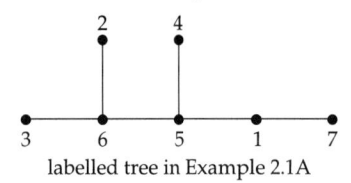

labelled tree in Example 2.1A

This happens in general — if you start with any labelled tree, find the corresponding Prüfer sequence, and then find the labelled tree corresponding to this sequence, you always get back to the original labelled tree. The two constructions above do indeed give us the required one–one correspondence between labelled trees and Prüfer sequences.

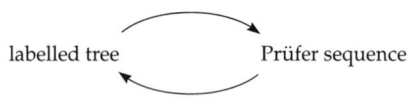

labelled tree ⇌ Prüfer sequence

We now show how this one–one correspondence can be used to prove Cayley's theorem.

Theorem 2.1: Cayley's theorem

The number of labelled trees with n vertices is n^{n-2}.

Proof

We assume that $n \geq 3$, since the result is clearly true if $n = 1$ or 2.

We construct the above one–one correspondence between the set of labelled trees with n vertices and the set of all Prüfer sequences of the form $(\mathbf{a_1}, \mathbf{a_2}, \mathbf{a_3}, \ldots, \mathbf{a_{n-2}})$, where each $\mathbf{a}_i$ is one of the integers $1, 2, 3, \ldots, n$ (allowing repetition). Since there are exactly n possible values for each integer $\mathbf{a}_i$, the total number of such sequences is

$$n \times n \times \cdots \times n = n^{n-2},$$
$$(n-2 \text{ terms})$$

and so, by the one–one correspondence, the number of labelled trees with n vertices is also n^{n-2}. ∎

Problem 2.5

Construct explicitly the one–one correspondence between the sixteen labelled trees with four vertices obtained in Problem 2.1 and the sixteen Prüfer sequences (a_1, a_2), where each term is 1, 2, 3 or 4.

Problem 2.6

How many irrigation canal systems are there linking eight locations with seven canals?

This question was posed at the beginning of Section 2.

Historical note

The earliest statement of Cayley's theorem occurred in his article *A theorem on trees*, written in 1889, although related results had been described earlier. However, Cayley's 'proof' was unsatisfactory, since he discussed only the case $n = 6$ and his argument cannot easily be generalized to larger values of n. Since then, several proofs have appeared, of which Prüfer's, given in 1918, is probably the best known.

2.2 Counting binary trees

In this section we illustrate a different technique — this time for counting unlabelled trees — by deriving an equation expressing the number of trees with a given number of vertices in terms of the number of trees with fewer vertices; such an equation is called a *recurrence relation*. Most problems of this kind are too complicated for us to consider here, but the following discussion of *binary trees* is intended to give you an idea of some of the techniques involved.

Definition

A **binary tree** is a rooted tree in which the number of edges coming down from each vertex is at most 2, and a distinction is made between left-hand and right-hand branches.

The binary trees with at most three vertices (including the root) are as follows; as before, we represent the root by a small square.

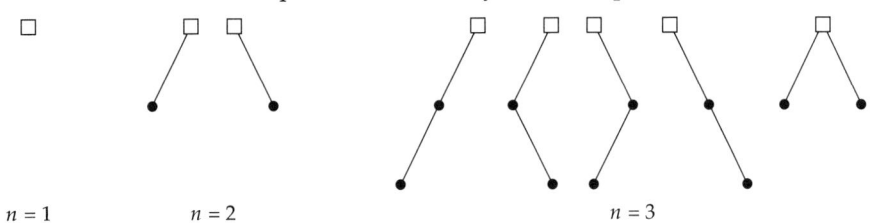

$n = 1$ $n = 2$ $n = 3$

Note that branching to the left and branching to the right at each stage give rise to *different* binary trees. Thus there are five *binary* trees with three vertices (whereas there are only two *rooted* trees with three vertices, as shown in the margin).

Problem 2.7

Draw the fourteen binary trees with four vertices.

We now consider the question: how many different binary trees are there with a given number of vertices?

Let u_n denote the number of binary trees with n vertices. Then, from the above diagrams and Problem 2.7, we have

$u_1 = 1$, $u_2 = 2$, $u_3 = 5$ and $u_4 = 14$.

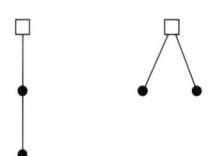

In order to find u_n, for a general value of n, we distinguish between those binary trees with one edge emerging from the root to the left, those with one edge emerging from the root to the right, and those with two edges emerging from the root. For example, for $n = 3$, the first two binary trees depicted above have a 'root edge' to the left, the next two have a 'root edge' to the right, and the last one has both.

Let a_n denote the number of binary trees with n vertices and a root edge to the left, let b_n denote the number of binary trees with n vertices and a root edge to the right, and let c_n denote the number of binary trees with n vertices and two root edges. Then, from the above diagrams and Problem 2.7, we have

for the binary tree with 1 vertex, $a_1 = 0$, $b_1 = 0$, $c_1 = 0$;

for the binary trees with 2 vertices, $a_2 = 1$, $b_2 = 1$, $c_2 = 0$;

for the binary trees with 3 vertices, $a_3 = 2$, $b_3 = 2$, $c_3 = 1$;

for the binary trees with 4 vertices, $a_4 = 5$, $b_4 = 5$, $c_4 = 4$;

in fact, for $n \geq 2$, we have

$$u_n = a_n + b_n + c_n.$$

Now consider a_n and b_n. Any binary tree with n vertices and a single root edge has one of the forms shown below, and can be obtained by taking a binary tree with $n - 1$ vertices rooted at Q, and joining it to the root R by the root edge RQ.

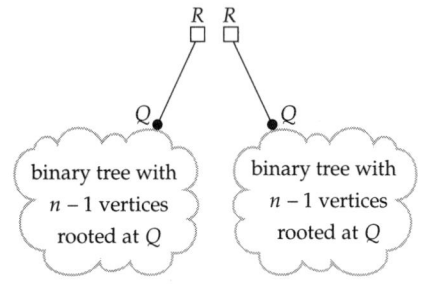

Since the number of binary trees rooted at Q is u_{n-1}, we deduce that

$$a_n = u_{n-1} \quad \text{and} \quad b_n = u_{n-1}, \quad \text{for } n \geq 2.$$

You can check from the above lists of numbers that this is correct for $n = 2$, 3, 4; for example, $a_4 = b_4 = u_3 = 5$.

Next consider c_n. Any binary tree with n vertices and two root edges has the form shown below, and can be obtained by taking a binary tree with k vertices rooted at P, and a binary tree with $(n-1) - k$ vertices rooted at Q, and joining them both to the root R by the root edges RP and RQ.

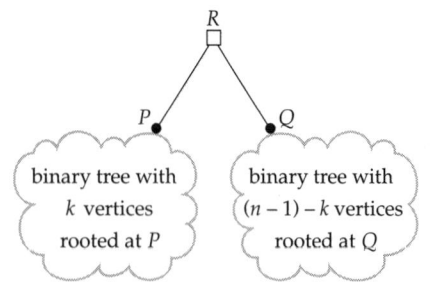

Since there are u_k such binary trees rooted at P, and u_{n-k-1} such binary trees rooted at Q, and as k can be any of the numbers $1, 2, 3, \ldots, n - 2$, we deduce that, for $n \geq 3$,

$$c_n = u_1 u_{n-2} + u_2 u_{n-3} + u_3 u_{n-4} + \ldots + u_{n-2} u_1.$$

Again, you can check from the above lists of numbers that this is correct for $n = 3, 4$; for example, $c_4 = u_1 u_2 + u_2 u_1 = (1 \times 2) + (2 \times 1) = 4$.

If we now substitute these expressions for a_n, b_n and c_n into the equation $u_n = a_n + b_n + c_n$, we obtain the following result.

$$u_n = 2u_{n-1} + (u_1 u_{n-2} + u_2 u_{n-3} + u_3 u_{n-4} + \ldots + u_{n-2} u_1)$$

Using this recurrence relation with $n = 5, 6, \ldots$, we can find the values of u_5, u_6, and so on, as far as we wish. For example,

$$u_5 = 2u_4 + (u_1 u_3 + u_2 u_2 + u_3 u_1)$$
$$= (2 \times 14) + (1 \times 5) + (2 \times 2) + (5 \times 1) = 42.$$

Problem 2.8

Use the above recurrence relation to determine the number of binary trees with six vertices.

2.3 Counting chemical trees

In *Graphs 1*, you saw how certain molecules can be represented as trees. In particular, we observed, but did not prove, that the graph of an alkane with formula $C_n H_{2n+2}$ is a tree. For example, the graphs of the three alkanes with formula $C_5 H_{12}$ are all trees:

pentane

2-methyl butane

2,2-dimethyl propane

Problem 2.9

Determine the numbers of vertices and edges in the graph of a molecule with formula $C_6 H_{14}$. Deduce that such a graph is a tree.

Using the list of unlabelled trees with six vertices, we can find all the alkanes with formula $C_6 H_{14}$. There are six such trees:

The unlabelled trees with six vertices are given in Solution 1.1.

The first five of these are the carbon graphs of the alkanes: hexane, 2-methyl pentane, 3-methyl pentane, 2,3-dimethyl butane and 2,2-dimethyl butane; the sixth tree cannot be the carbon graph of a molecule, as it has a vertex of degree 5.

We now ask you to show that the graph of any alkane is a tree.

Problem 2.10

By determining the numbers of vertices and edges, deduce that the graph of any alkane with formula C_nH_{2n+2} is a tree.

The general problem of determining the number of alkanes with formula C_nH_{2n+2} was solved in the 1870s by the English mathematician Arthur Cayley. In order to outline his method, we first need to introduce the concept of the 'middle' of a tree. For some trees this is easy to define:

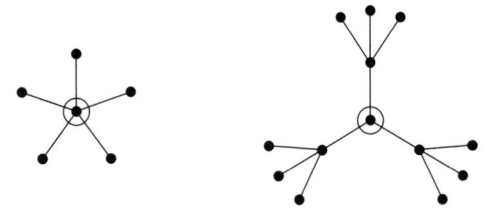

But how do we define the 'middle' of the following trees?

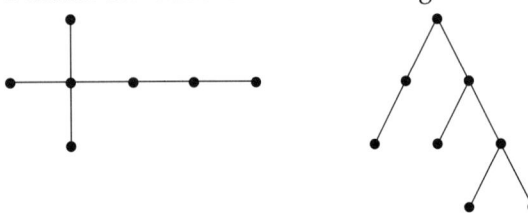

To answer this question, we take our tree and remove all the vertices of degree 1, together with their incident edges; we then repeat this process until we obtain *either* a single vertex, called the **centre**, *or* two vertices joined by an edge, called the **bicentre**. A tree with a centre is called a **central tree**, and a tree with a bicentre is called a **bicentral tree**; every tree is either central or bicentral, but not both. For example, the following tree is a *central tree* with centre e.

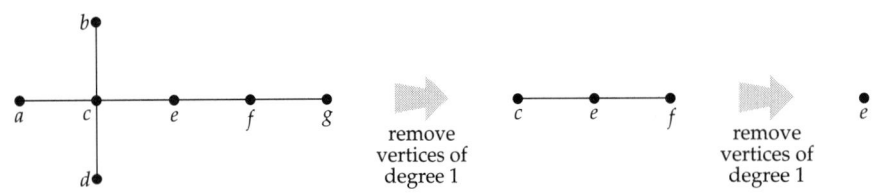

The following tree is a *bicentral tree* with bicentre cd.

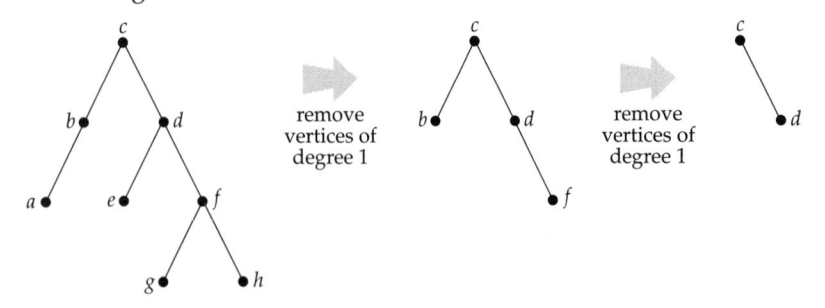

Problem 2.11

Classify each of the trees with five and six vertices as *central* or *bicentral*, and locate the centre or bicentre in each case.

The unlabelled trees with five vertices are shown at the beginning of Section 1.1, and those with six vertices are shown above.

Cayley's approach to the problem of finding the number of alkanes was to regard each molecule as a rooted tree in which the root is the centre (a carbon vertex of degree 4) or the bicentre (two such vertices, joined by an edge). By removing this root, he produced a number of smaller rooted trees; by this means he obtained a recurrence relation, somewhat similar to the one we derived for binary trees in Section 2.2, that successively gives the number of alkanes with formula C_nH_{2n+2}, for increasing values of n. Although his recurrence relation was extremely complicated, Cayley was able to use it to calculate correctly the number of alkanes with up to eleven carbon atoms.

Historical note

Arthur Cayley had been interested in trees for almost twenty years before solving the alkane-counting problem. His first two papers on trees, entitled *On the theory of the analytical forms called trees*, had appeared in 1857 and 1859, and were concerned with the problem of counting rooted trees in connection with a problem arising from the differential calculus.

ARTHUR CAYLEY (1821–95)

2.4 Postscript

Many difficult enumeration problems can be solved using a result due to Georg Pólya, known as *Pólya's counting theorem*, proved in the 1930s. Among the objects which can be counted in this way are the unlabelled graphs with a given number of vertices. Some of Pólya's work was anticipated by Howard Redfield in the late 1920s. Unfortunately, Redfield's work was written in complicated language, and had little influence on the later development of the subject.

In recent years, many kinds of graph have been counted by means of Pólya's theorem, as described in a paper by Frank Harary.

Note that in counting the number of a certain set of objects, we have sometimes attacked only half the enumeration problem. We might be interested in *listing* the actual objects we have counted. The computer comes into its own here, and computer algorithms can be used to generate the objects of interest. Major contributions to this aspect of enumeration have been made by the mathematician Ronald Read, who has developed many elegant and efficient algorithms.

Pólya had a theorem
    (Which Redfield proved of old).
What secrets sought by graphmen
    Were by the theorem told!

Thus Pólya counted finite trees
    (As Redfield did before).
'Their number is exactly such,
    And not a seedling more.'

Harary counted finite graphs
    (Like Redfield long ago).
And pointed out how very much
    To Pólya's work we owe.

And Read piled graph on graph on graph
    (Which is what Redfield did).
So numbering the graphic world
    That nothing could be hid.

Then hail, Harary, Pólya, Read,
    Who taught us graphic lore!
(And spare a thought for Redfield too
    Who went too long before).

BLANCHE DESCARTES

After studying this section, you should be able to:

- find the *Prüfer sequence* associated with a given labelled tree, and vice versa;
- state and use Cayley's theorem for labelled trees;
- understand the method for counting binary trees;
- distinguish between *central* and *bicentral* trees, and describe their use in the counting of alkanes.

3 Greedy algorithms

In this section we turn our attention to some common algorithms associated with the construction of trees. In particular, we return to the *minimum connector problem*, and show how its solution can give us information about the solution of the *travelling salesman problem*. Both of these topics involve the idea of *spanning trees* in a graph, and we start with a brief subsection to introduce this idea.

These problems were discussed in the *Introduction* unit.

3.1 Spanning trees

> **Definition**
>
> Let G be a connected graph. Then a **spanning tree** in G is a subgraph of G that includes every vertex and is also a tree.

For example, the following diagram shows a graph and three of its spanning trees.

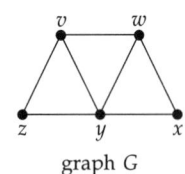

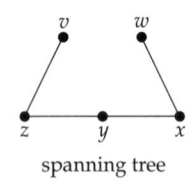

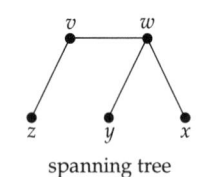

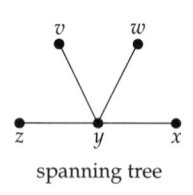

The number of spanning trees in a graph can be very large; for example, the Petersen graph has 2000 labelled spanning trees.

Problem 3.1 ─────────────

The above graph G has twenty-one spanning trees. Find as many of them as you can.

Problem 3.2 ─────────────

Find three spanning trees in the Petersen graph (shown in the margin).

Problem 3.3 ─────────────

Use Cayley's theorem from Section 2.1 to write down the number of labelled spanning trees in the complete graph K_n.

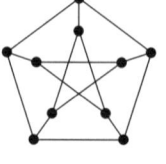

Given a connected graph, we can construct a spanning tree by using either of the following two methods. Each method can be extended to the construction of a *minimum-weight spanning tree* in a weighted graph, as you will see in the next subsection.

Building-up method
Select edges of the graph one at a time, in such a way that no cycles are created;
repeat this procedure until all vertices are included.

Example
In the above graph G, we choose the edges vz, wx, xy and yz; then no cycles are created, and we obtain the first of the above spanning trees.

Cutting-down method
Choose any cycle and remove any one of its edges;
repeat this procedure until no cycles are left.

Example
From the above graph G, we remove the edges
vy (destroying the cycle $vwyv$),
yz (destroying the cycle $vwyzv$),
xy (destroying the cycle $wxyw$),
and we obtain the second of the above spanning trees.

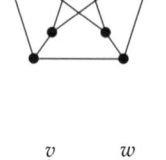

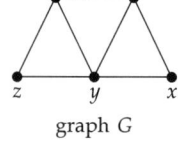

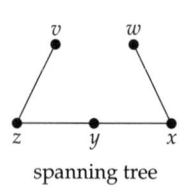

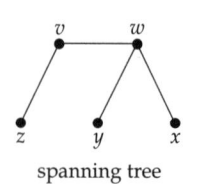

26

Problem 3.4

Use each method to construct a spanning tree in the complete graph K_5.

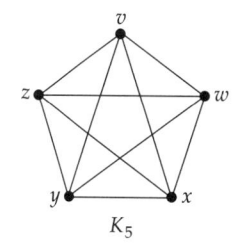

3.2 Minimum connector problem

Consider the following problem.

An irrigation canal system is to be built, interconnecting a number of given locations. The cost of digging and maintaining each canal in the system is known. Some pairs of locations cannot be joined by a canal for geographical or political reasons (for example, there is a gorge or an environmentally sensitive area). How do we design a canal system which interconnects all the locations at minimum total cost?

The minimum connector problem was introduced in the *Introduction* unit and is the main topic of the article in the Appendix.

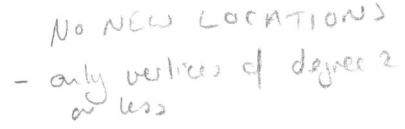

This problem can be interpreted in two ways, depending on whether or not we allow extra 'locations' where canals may intersect. For example, in the case of the canal system shown below, we may be able to reduce the total cost by creating an extra location at the point E, and linking it to A.

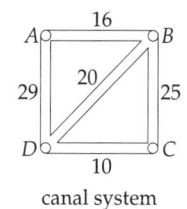

canal system

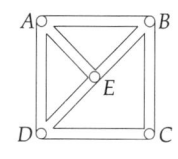

system with extra location E

Unfortunately, for many such problems, the additional cost of inserting an extra location (which may be a telephone exchange or power station) can greatly exceed the possible saving in cost from a shorter connection system, and the resulting mathematical analysis becomes rather complicated. In view of this, we adopt the second interpretation of the problem and assume that each connection joins only two existing locations; no new locations are allowed.

NO NEW LOCATIONS
— only vertices of degree 2 or less

The insertion of extra locations is discussed in the article in the Appendix.

We can model this problem graphically by representing the locations by vertices and the canals by edges, giving us a weighted graph in which the weights are the costs. The problem is then to find a subgraph of minimum total weight, passing through each vertex. Such a subgraph is necessarily a spanning tree, because, if there is a cycle, we can lower the total cost by removing any one of its edges.

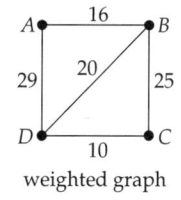

weighted graph

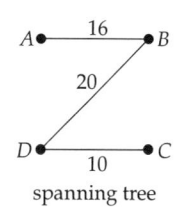

spanning tree

In our example, the graph has total weight 100. The removal of one edge from the cycle $ABDA$ and an edge from the remaining cycle — $ABCDA$ or $BCDB$ — lowers the total weight, and gives us a spanning tree. The spanning tree of minimum total weight is obtained by removing the two edges of maximum weight — AD and BC. The minimum total cost is thus $16 + 20 + 10 = 46$. We call this spanning tree a *minimum connector* for the locations $A, \ldots, D$.

Here we use the *cutting-down* method for obtaining a spanning tree.

Definition

Let T be a spanning tree of minimum total weight in a connected weighted graph G. Then T is a **minimum spanning tree** or a **minimum connector** in G.

We now restate the minimum connector problem in graphical terms.

Minimum connector problem

Given a weighted graph, find a minimum spanning tree.

In the *Introduction* unit we presented two algorithms for solving the minimum connector problem. Both belong to a class of algorithms known as *greedy algorithms*. The name arises from the fact that at each stage we make the 'greediest', or best, choice available, regardless of the subsequent effect of that choice. Such 'local' algorithms do not succeed for most combinatorial problems, but this is one case where they do. Each algorithm was presented formally as a sequence of steps, and then summarized informally in a form suitable for small examples to be tackled without the use of a computer. Here we use similar informal descriptions of the algorithms. The first algorithm we gave was *Kruskal's algorithm*.

Kruskal's greedy algorithm for a minimum connector

To construct a minimum spanning tree in a connected weighted graph G, successively choose edges of G of minimum weight in such a way that no cycles are created, until a spanning tree is obtained.

This algorithm was derived by Joseph Kruskal in 1956, but had been stated in 1926 by O. Borůvka.

We illustrate the use of this algorithm in the following example.

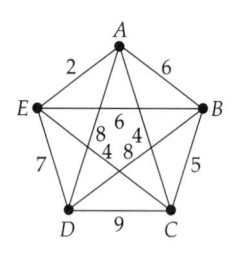

Example 3.1

Consider the given weighted graph.

We apply Kruskal's algorithm, successively choosing edges of minimum weight in such a way that no cycles are created.

Note that, in this example, some of the weights are the same, so we may have a choice of edges at some stages, and there may be more than one minimum spanning tree. (This situation did not arise in the examples considered in the *Introduction* unit.)

First edge

We choose an edge of minimum weight — the only possibility is *AE*, with weight 2.

Second edge

We choose an edge of next smallest weight — we can choose either *AC* or *CE*, with weight 4; let us choose *CE*.

Third edge

We cannot now choose *AC*, also with weight 4, as this would create a cycle (*ACEA*), so we choose an edge of next smallest weight — the only possibility is *BC*, with weight 5.

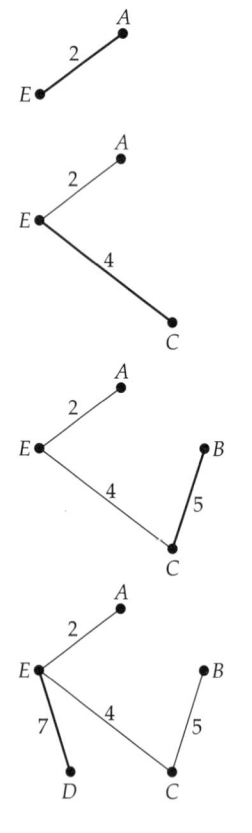

Fourth edge

The edges of next smallest weight are *AB* and *BE*, with weight 6. We cannot choose either of these, as each would create a cycle (*ABCEA* or *BCEB*), so we choose an edge of next smallest weight — the only possibility is *DE*, with weight 7.

This completes a spanning tree, which is a minimum spanning tree of total weight $2 + 4 + 5 + 7 = 18$. ∎

Problem 3.5

If we had chosen the edge AC at the second stage in Example 3.1, rather than the edge CE, which minimum spanning tree would we have obtained?

When there are several vertices in a weighted graph, the weights are usually given in tabular form, as illustrated by the following example.

Example 3.2

The following table gives the distances (in hundreds of miles) between six cities. We use Kruskal's algorithm to find a minimum spanning tree connecting these cities.

	Berlin	London	Moscow	Paris	Rome	Seville
Berlin	–	7	11	7	10	15
London	7	–	18	3	12	11
Moscow	11	18	–	18	20	27
Paris	7	3	18	–	9	8
Rome	10	12	20	9	–	13
Seville	15	11	27	8	13	–

If you wish to draw a diagram for this situation, draw the complete graph K_6 with the six vertices corresponding to the cities.

For a small table such as this, we can select the distances in ascending order directly from the table. However, you may prefer to begin by drawing up a table of the distances in ascending order as in the *Introduction* unit.

We apply Kruskal's algorithm, successively choosing edges of minimum weight in such a way that no cycles are created.

First edge

We choose an edge of minimum weight — the only possibility is London–Paris, with weight 3.

Second edge

We choose an edge of next smallest weight — we can choose either Berlin–London or Berlin–Paris, with weight 7; let us choose Berlin–London.

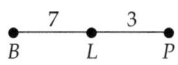

Third edge

We cannot now choose Berlin–Paris, also with weight 7, as this would create a cycle, so we choose an edge of next smallest weight — the only possibility is Paris–Seville, with weight 8.

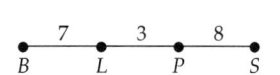

Fourth edge

We choose an edge of next smallest weight — the only possibility is Paris–Rome, with weight 9.

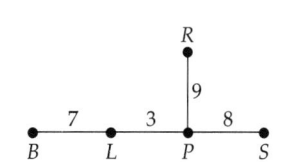

Fifth edge

We cannot now choose Berlin–Rome (weight 10) or London–Seville (weight 11), as each would create a cycle, so we next choose Berlin–Moscow, with weight 11.

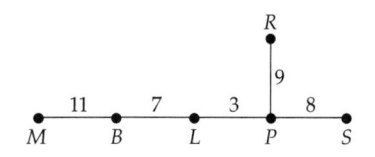

This completes a spanning tree, which is a minimum spanning tree of total weight $3 + 7 + 8 + 9 + 11 = 38$. ∎

Problem 3.6

The following table gives the distances (in miles) between six places in Ireland. Use Kruskal's algorithm to find a minimum spanning tree connecting these places.

	Athlone	Dublin	Galway	Limerick	Sligo	Wexford
Athlone	–	78	56	73	71	114
Dublin	78	–	132	121	135	96
Galway	56	132	–	64	85	154
Limerick	73	121	64	–	144	116
Sligo	71	135	85	144	–	185
Wexford	114	96	154	116	185	–

Although Kruskal's algorithm can easily be applied without the use of a computer when the graph is small, it is not particularly well suited to efficient computer implementation, owing to the need to arrange the edges in order of ascending weight, and the need to recognize cycles as they are created. Both of these difficulties can be overcome easily by a slight modification of the algorithm; the result is known as *Prim's algorithm*. Both algorithms are polynomial-time algorithms.

Prim's greedy algorithm for a minimum connector

To construct a minimum spanning tree T in a connected weighted graph G, build up T step by step as follows:

- put an arbitrary vertex from the graph G into T;
- successively add edges of minimum weight joining a vertex already in T to a vertex not in T, until a spanning tree is obtained.

Greedy algorithms and polynomial-time algorithms are discussed in the article in the Appendix.

Note that with Prim's algorithm we obtain a *connected* graph at each stage.

Example 3.3

The following table gives the distances (in hundreds of miles) between six cities. We use Prim's algorithm to find a minimum spanning tree connecting these cities:

	Berlin	London	Moscow	Paris	Rome	Seville
Berlin	–	7	11	7	10	15
London	7	–	18	3	12	11
Moscow	11	18	–	18	20	27
Paris	7	3	18	–	9	8
Rome	10	12	20	9	–	13
Seville	15	11	27	8	13	–

This is the data we used in Example 3.2 with Kruskal's algorithm.

We apply Prim's algorithm, successively choosing edges of minimum weight joining a vertex already in T to a vertex not in T.

We start by choosing any vertex and putting it in T; let us choose Berlin.

First edge

We choose an edge of minimum weight joining Berlin to another vertex; we can choose either Berlin–London or Berlin–Paris, with weight 7; let us choose Berlin–London.

Second edge

We choose an edge of minimum weight joining Berlin or London to another vertex; the only possibility is London–Paris, with weight 3.

Third edge

We choose an edge of minimum weight joining Berlin, London or Paris to another vertex; the only possibility is Paris–Seville, with weight 8.

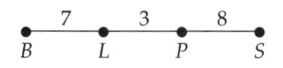

Fourth edge

We choose an edge of minimum weight joining Berlin, London, Paris or Seville to another vertex; the only possibility is Paris–Rome, with weight 9.

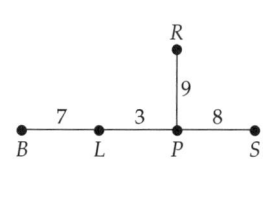

Fifth edge

We choose an edge of minimum weight joining Berlin, London, Paris, Rome or Seville to the remaining vertex; the only possibility is Berlin–Moscow, with weight 11.

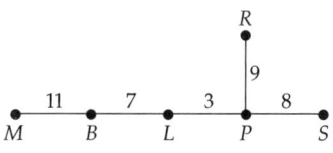

This completes the spanning tree, which is a minimum spanning tree of weight $7 + 3 + 8 + 9 + 11 = 38$. It is the tree we obtained in Example 3.2. ∎

Problem 3.7

In the above example, which minimum spanning tree would we have obtained

(a) if at the first stage we had chosen Berlin–Paris, instead of Berlin–London?

(b) if we had started by choosing Rome as our first vertex, instead of Berlin?

A convenient method of applying Prim's algorithm is the following tabular or matrix method, which starts from the vertex corresponding to row 1 and column 1 in the table of weights. The method systematically identifies an edge in the minimum connector (identified by circling the corresponding weight in the matrix) and reduces the matrix by removing the remaining entries in the appropriate column.

Here we treat the table of weights as a matrix and label the rows and columns accordingly.

We denote the weight in row i and column j of the matrix by w_{ij}.

Tabular (matrix) form of Prim's greedy algorithm for a minimum connector

START   with a table of weights for a connected weighted graph, and with no circled entries in the matrix.

STEP 1   Delete all entries in column 1, and mark row 1 with a star.

STEP 2   Select a smallest entry from the uncircled entries in the row(s) marked with a star.

If no such entry exists, STOP: the edges corresponding to the circled weights form a minimum connector, and its total weight is the sum of the circled weights.

Otherwise, go on to Step 3.

STEP 3   (a) Circle the weight w_{ij} identified in Step 2.
(b) Mark row j with a star.
(c) Delete the remaining entries in column j.

Return to Step 2.

For convenience, we start with row 1 and column 1, but it is not necessary to start here.

We illustrate the use of the algorithm by applying it to the European cities table in Examples 3.2 and 3.3.

Example 3.4

We start with the matrix:

$$\begin{array}{c} & \begin{array}{cccccc} 1 & 2 & 3 & 4 & 5 & 6 \end{array} \\ \begin{array}{c} 1 \\ 2 \\ 3 \\ 4 \\ 5 \\ 6 \end{array} & \left[\begin{array}{cccccc} - & 7 & 11 & 7 & 10 & 15 \\ 7 & - & 18 & 3 & 12 & 11 \\ 11 & 18 & - & 18 & 20 & 27 \\ 7 & 3 & 18 & - & 9 & 8 \\ 10 & 12 & 20 & 9 & - & 13 \\ 15 & 11 & 27 & 8 & 13 & - \end{array} \right] \end{array}$$

STEP 1 We delete all entries in column 1, and mark row 1 with a star.

$$\begin{array}{c} & \begin{array}{cccccc} 1 & 2 & 3 & 4 & 5 & 6 \end{array} \\ \begin{array}{c} 1 \\ 2 \\ 3 \\ 4 \\ 5 \\ 6 \end{array} & \left[\begin{array}{cccccc} - & 7 & 11 & 7 & 10 & 15 \\ - & - & 18 & 3 & 12 & 11 \\ - & 18 & - & 18 & 20 & 27 \\ - & 3 & 18 & - & 9 & 8 \\ - & 12 & 20 & 9 & - & 13 \\ - & 11 & 27 & 8 & 13 & - \end{array} \right] \begin{array}{c} * \\ \\ \\ \\ \\ \end{array} \end{array}$$

In practice, we do not repeat the table as in this example, but work on a single copy of the table, deleting entries by crossing through them.

First iteration

STEP 2 The smallest uncircled entry in the starred row is 7 in columns 2 and 4; we select 7 in column 2.

STEP 3 We circle the 7 in row 1 and column 2, mark row 2 with a star, and delete the remaining entries in column 2.

$$\begin{array}{c} & \begin{array}{cccccc} 1 & 2 & 3 & 4 & 5 & 6 \end{array} \\ \begin{array}{c} 1 \\ 2 \\ 3 \\ 4 \\ 5 \\ 6 \end{array} & \left[\begin{array}{cccccc} - & ⑦ & 11 & 7 & 10 & 15 \\ - & - & 18 & 3 & 12 & 11 \\ - & - & - & 18 & 20 & 27 \\ - & - & 18 & - & 9 & 8 \\ - & - & 20 & 9 & - & 13 \\ - & - & 27 & 8 & 13 & - \end{array} \right] \begin{array}{c} * \\ * \\ \\ \\ \\ \end{array} \end{array}$$

Second iteration

STEP 2 The smallest uncircled entry in the starred rows is 3 in row 2 and column 4, so we select 3.

STEP 3 We circle 3 in row 2 and column 4, mark row 4 with a star, and delete the remaining entries in column 4.

$$\begin{array}{c} & \begin{array}{cccccc} 1 & 2 & 3 & 4 & 5 & 6 \end{array} \\ \begin{array}{c} 1 \\ 2 \\ 3 \\ 4 \\ 5 \\ 6 \end{array} & \left[\begin{array}{cccccc} - & ⑦ & 11 & - & 10 & 15 \\ - & - & 18 & ③ & 12 & 11 \\ - & - & - & - & 20 & 27 \\ - & - & 18 & - & 9 & 8 \\ - & - & 20 & - & - & 13 \\ - & - & 27 & - & 13 & - \end{array} \right] \begin{array}{c} * \\ * \\ \\ * \\ \\ \end{array} \end{array}$$

Third iteration

STEP 2 The smallest uncircled entry in the starred rows is 8 in row 4 and column 6, so we select 8.

STEP 3 We circle 8 in row 4 and column 6, mark row 6 with a star, and delete the remaining entries in column 6.

$$\begin{array}{c} & \begin{array}{cccccc} 1 & 2 & 3 & 4 & 5 & 6 \end{array} \\ \begin{array}{c} 1 \\ 2 \\ 3 \\ 4 \\ 5 \\ 6 \end{array} & \left[\begin{array}{cccccc} - & ⑦ & 11 & - & 10 & - \\ - & - & 18 & ③ & 12 & - \\ - & - & - & - & 20 & - \\ - & - & 18 & - & 9 & ⑧ \\ - & - & 20 & - & - & - \\ - & - & 27 & - & 13 & - \end{array} \right] \begin{array}{c} * \\ * \\ \\ * \\ \\ * \end{array} \end{array}$$

Fourth iteration

STEP 2 The smallest uncircled entry in the starred rows is 9 in row 4 and column 5, so we select 9.

STEP 3 We circle 9 in row 4 and column 5, mark row 5 with a star, and delete the remaining entries in column 5.

$$\begin{array}{c} & \begin{array}{cccccc} 1 & 2 & 3 & 4 & 5 & 6 \end{array} \\ \begin{array}{c} 1 \\ 2 \\ 3 \\ 4 \\ 5 \\ 6 \end{array} & \left[\begin{array}{cccccc} - & ⑦ & 11 & - & - & - \\ - & - & 18 & ③ & - & - \\ - & - & - & - & - & - \\ - & - & 18 & - & ⑨ & ⑧ \\ - & - & 20 & - & - & - \\ - & - & 27 & - & - & - \end{array} \right] \begin{array}{c} * \\ * \\ \\ * \\ * \\ * \end{array} \end{array}$$

32

Fifth iteration

STEP 2 The smallest uncircled entry in the starred rows is 11 in row 1 and column 3, so we select 11.

STEP 3 We circle 11 in row 1 and column 3, mark row 3 with a star, and delete the remaining entries in column 3.

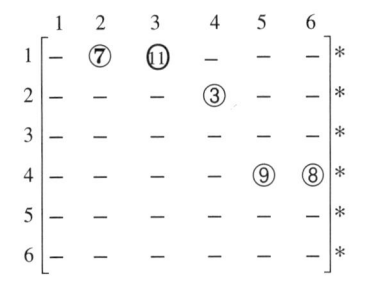

$$\begin{array}{c} & 1 & 2 & 3 & 4 & 5 & 6 \\ 1 & - & ⑦ & ⑪ & - & - & - \\ 2 & - & - & - & ③ & - & - \\ 3 & - & - & - & - & - & - \\ 4 & - & - & - & - & ⑨ & ⑧ \\ 5 & - & - & - & - & - & - \\ 6 & - & - & - & - & - & - \end{array} \begin{array}{c} * \\ * \\ * \\ * \\ * \\ * \end{array}$$

Sixth iteration

STEP 2 No uncircled numbers exist in the starred rows, so we STOP.

The minimum connector has length 7 + 11 + 3 + 9 + 8 = 38, and includes the edges Berlin–London, Berlin–Moscow, London–Paris, Paris–Rome and Paris–Seville.

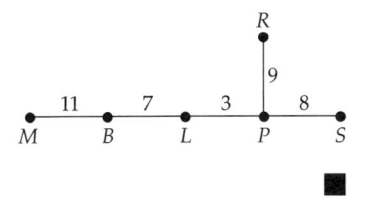

This is the tree we obtained in Examples 3.2 and 3.3.

∎

Problem 3.8

The following table gives the distances (in miles) between five villages A, B, C, D and E.

	A	B	C	D	E
A	–	6	4	8	2
B	6	–	5	8	6
C	4	5	–	9	4
D	8	8	9	–	7
E	2	6	4	7	–

Use the tabular (matrix) form of Prim's algorithm, starting at A, to find a minimum connector for the five villages.

Comparison of Kruskal's and Prim's algorithms

The execution time of Prim's algorithm depends only on the number of vertices, but the time for Kruskal's algorithm for a graph with the same number of vertices increases as the number of edges is increased. However, in general, it is not possible to assert which one is more efficient. The efficiency depends on, among other things, the structure of the graph and the distribution of weights. It has been observed that for graphs with up to 100 vertices, Prim's method appears to be more efficient, particularly so when there is an abundance of edges. The following running times for the two algorithms run on an AMDHL 470 V/8 computer are reported by Syslo et al. (1983).

number of vertices	number of edges	execution time (in msec) Kruskal	Prim
80	790	64	52
80	1580	95	51
80	3160	186	50
100	200	44	78
100	300	66	80

We close this subsection by outlining a proof of the fact that Prim's and Kruskal's algorithms, as stated above, always produce a minimum connector.

Theorem 3.1

Prim's and Kruskal's algorithms always produce a spanning tree of minimum weight.

Outline of proof

Suppose that the algorithm produces a tree T, and that there exists a spanning tree S with smaller total weight than T. Let e be an edge of smallest weight lying in T but not in S.

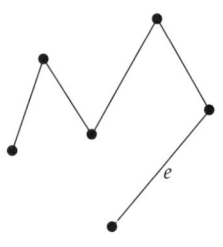

tree T

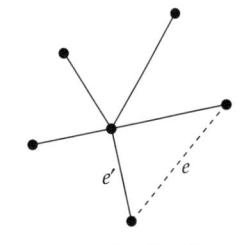

spanning tree S
(2 edges in common with T)

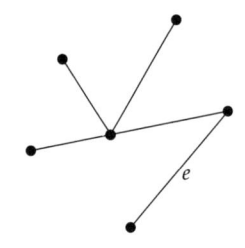

spanning tree S'
(3 edges in common with T)

If we add the edge e to S we create a cycle containing e. Since this cycle must contain an edge e' not contained in T, the subgraph obtained from S on replacing e' by e is also a spanning tree (S', say). It follows from the construction of T that the weight of e cannot exceed the weight of e'. So the total weight of S' does not exceed the total weight of S, and S' has one more edge in common with T than has S.

It follows, on repeating this procedure, that we can change S into T, one edge at a time, with the weight not increasing at each stage. This shows that the total weight of T does not exceed the total weight of S, contradicting the definition of S. This contradiction establishes the result. ∎

3.3 Maximum connector problem

Up to now, our concern has been to find a spanning tree of *minimum* weight in a given graph. However, both the algorithms of Kruskal and Prim can be modified to give a spanning tree of *maximum* weight; such a tree is called a **maximum spanning tree** or **maximum connector**.

We now state the *maximum connector problem* in graphical terms, as follows.

Maximum connector problem

Given a weighted graph, find a maximum spanning tree.

The following is the version of Prim's algorithm for the determination of a maximum connector.

We use the version of Kruskal's algorithm in Section 4.1.

Prim's greedy algorithm for a maximum connector

To construct a maximum spanning tree T in a connected weighted graph G, build up T step by step as follows:

- put an arbitrary vertex from the graph G into T;
- successively add edges of maximum weight joining a vertex already in T to a vertex not in T, until a spanning tree is obtained.

We illustrate the use of this algorithm by the following example.

Example 3.5

The following tables gives the distances (in hundreds of miles) between six cities. We use Prim's algorithm to find a maximum spanning tree connecting these cities:

This is the data used in Examples 3.2–3.4.

	Berlin	London	Moscow	Paris	Rome	Seville
Berlin	–	7	11	7	10	15
London	7	–	18	3	12	11
Moscow	11	18	–	18	20	27
Paris	7	3	18	–	9	8
Rome	10	12	20	9	–	13
Seville	15	11	27	8	13	–

We apply Prim's algorithm, successively choosing edges of maximum weight joining a vertex already in T to a vertex not in T.

We start by choosing any vertex and putting it in T; let us choose Berlin.

First edge

We choose an edge of maximum weight joining Berlin to another vertex; the only possibility is Berlin–Seville, with weight 15.

Second edge

We choose an edge of maximum weight joining Berlin or Seville to another vertex; the only possibility is Seville–Moscow, with weight 27.

Third edge

We choose an edge of maximum weight joining Berlin, Seville or Moscow to another vertex; the only possibility is Moscow–Rome, with weight 20.

Fourth edge

We choose an edge of maximum weight joining Berlin, Seville, Moscow or Rome to another vertex; there are two possibilities: Moscow–London and Moscow–Paris, with weight 18. Let us choose Moscow–London.

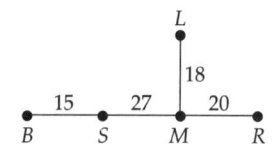

Fifth edge

We choose an edge of maximum weight joining Berlin, Seville, Moscow, Rome or London to the remaining vertex; the only possibility is Moscow–Paris, with weight 18.

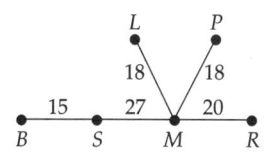

This completes the spanning tree, which is a maximum spanning tree of weight 15 + 27 + 20 + 18 + 18 = 98. ∎

Problem 3.9

The following table gives the distances (in miles) between six places in Ireland. Use Prim's algorithm, starting with Athlone, to find a *maximum* spanning tree connecting these places.

You found a minimum connector for these places in Problem 3.6.

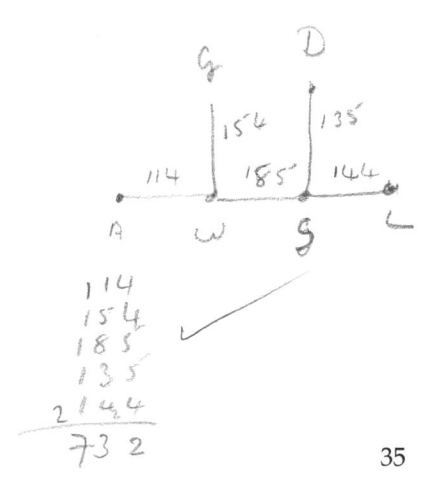

	Athlone	Dublin	Galway	Limerick	Sligo	Wexford
Athlone	–	78	56	73	71	114
Dublin	78	–	132	121	135	96
Galway	56	132	–	64	85	154
Limerick	73	121	64	–	144	116
Sligo	71	135	85	144	–	185
Wexford	114	96	154	116	185	–

3.4 Travelling salesman problem

Another problem you met in the *Introduction* unit is the *travelling salesman problem*, in which a salesman wishes to visit a number of cities and then return to the starting point, covering the *minimum* possible total distance. This is an important problem in practice, and can appear in a number of different guises; for example, the salesman may be concerned to minimize not the total distance but the total time taken, or the total cost.

We can express the problem in graphical language. In such terms, we are given a weighted graph, usually a complete graph, and our aim is to find a cycle of minimum total weight passing through every vertex — in other words, a *minimum-weight Hamiltonian cycle*.

> **Travelling salesman problem**
>
> Given a weighted complete graph, find a minimum-weight Hamiltonian cycle.

In view of the simple nature of Kruskal's and Prim's algorithms for solving the minimum connector problem, we might hope that there is a simple algorithm here as well. Unfortunately, no such algorithm is known. We could, of course, try all possible Hamiltonian cycles, but this is a hopeless task, even on a computer, unless the number of vertices is very small. For a travelling salesman problem involving 100 cities, there would be 100! ($\approx 9.3 \times 10^{157}$) sequences to be considered, which is way beyond present computer capacities. In fact, there is no known efficient algorithm for the travelling salesman problem — it is an NP-complete problem.

In view of this, we are forced to look for *approximate solutions* to the problem. One method for finding an approximate solution to the travelling salesman problem, which often works well in practice, is to find a *lower bound* for the total weight of a minimum-weight Hamiltonian cycle by solving a related minimum connector problem instead. We then know that the correct solution must exceed this lower bound.

We introduced the idea of an NP-complete problem in the *Introduction* unit, Section 4. See also the article in the Appendix.

We use the phrase 'solution to the travelling salesman problem' to mean the *total weight of a minimum-weight Hamiltonian cycle*. We are finding the length (*weight*) of the cycle — *we are not finding a cycle*.

We illustrate the method by the following example.

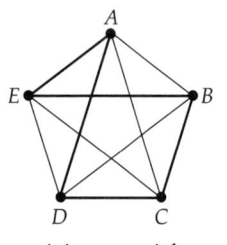

minimum–weight
Hamiltonian cycle
ADCBEA

If we take a minimum-weight Hamiltonian cycle in this (weighted) complete graph and remove the vertex A and its incident edges, we get a path passing through the remaining vertices.

The *actual* weights are irrelevant here, so they are not shown.

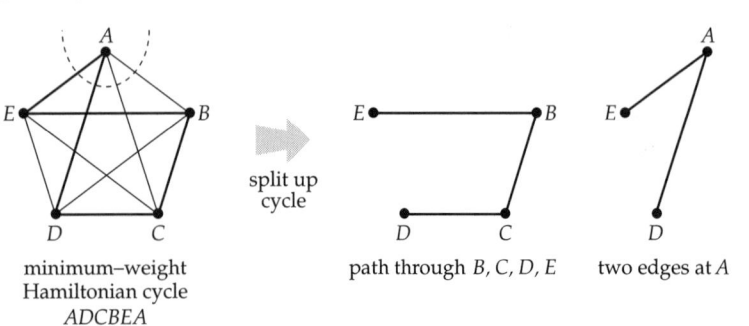

minimum–weight
Hamiltonian cycle
ADCBEA

split up cycle

path through B, C, D, E

two edges at A

Such a path must be a spanning tree for the complete graph formed by these remaining vertices, and the weight of the Hamiltonian cycle is obtained by adding the weight of this spanning tree to the weights of the two edges incident to A:

| total weight of minimum-weight Hamiltonian cycle | = | total weight of spanning tree connecting B, C, D, E | + | sum of weights of two edges incident to A |

It follows that:

| total weight of minimum-weight Hamiltonian cycle | $\geq$ | total weight of *minimum* spanning tree connecting B, C, D, E | + | sum of two *smallest* weights of edges incident to A |

So the expression on the right-hand side gives us a *lower bound* for the solution to the travelling salesman problem in this case.

In general, we can use the generalized form of this inequality to obtain a lower bound to any travelling salesman problem, as outlined below.

Method for finding a lower bound for the solution to the travelling salesman problem

STEP 1   Choose a vertex V and remove it from the graph.

STEP 2   Find a minimum spanning tree connecting the remaining vertices, and calculate its total weight W.

STEP 3   Find the two smallest weights, w_1 and w_2, of the edges incident to V.

STEP 4   Calculate the lower bound $W + w_1 + w_2$.

Different choices of V give *different* lower bounds. The *best* lower bound is the *largest*: it gives the most information about the actual solution.

Use Kruskal's or Prim's algorithm.

If you are saving up for a new hat, then it is more helpful to know that you need at least £30 than to know that you need at least £20.

Example 3.6

Consider the following weighted graph.

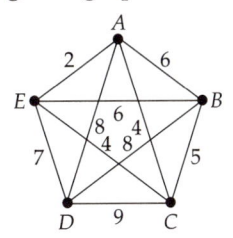

If we remove the vertex A, then the remaining weighted graph has the four vertices B, C, D and E.

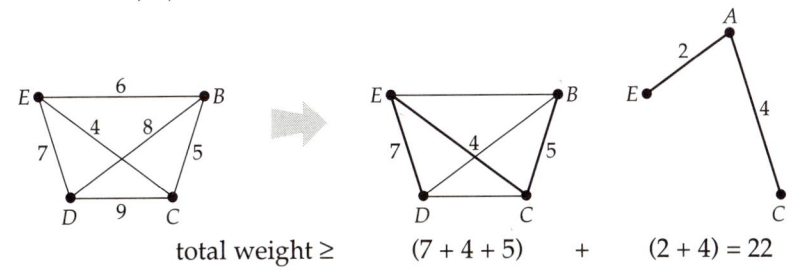

total weight $\geq$ $(7 + 4 + 5)$ + $(2 + 4) = 22$

The minimum spanning tree joining these vertices is the tree whose edges are BC, CE and ED, with total weight $5 + 4 + 7 = 16$. The two edges of smallest weight incident to A are AE and AC, with total weight $2 + 4 = 6$. The required lower bound for the solution to this instance of the travelling salesman problem is therefore $16 + 6 = 22$.

A little experimentation will show you that this lower bound is not very good: the actual solution to this problem is the cycle $ACBDEA$ with total weight 26, so our lower bound is not the correct solution.

A better lower bound (that is, one giving more information) can be obtained by removing the vertex D instead of A. In this case, the remaining weighted graph has the four vertices A, B, C and E, and there are two minimum spanning trees joining these vertices, each with total weight 11 (only one is highlighted below).

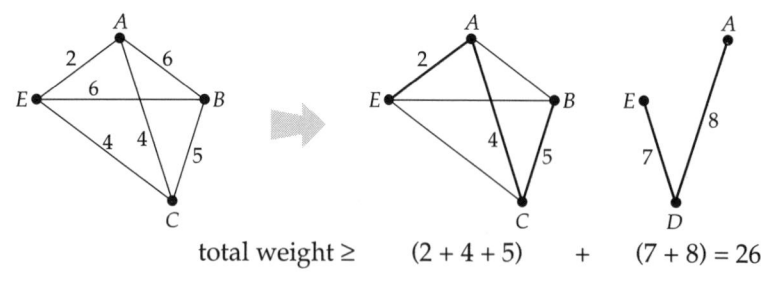

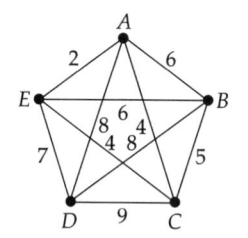

$$\text{total weight} \geq \quad (2 + 4 + 5) \quad + \quad (7 + 8) = 26$$

The two edges of smallest weight incident to D are DE, and DA or DB, with total weight 15. A better lower bound for the solution to the travelling salesman problem is therefore $11 + 15 = 26$, which is the correct solution. ∎

In the above example, we found a minimum spanning tree by inspection of the diagram (with the given vertex removed). However, for larger examples, a diagram becomes congested and it is easier to find a minimum spanning tree using Kruskal's or Prim's algorithm.

Problem 3.10

For Example 3.6, find the lower bounds obtained by removing:

(a) vertex B; (b) vertex E.

Which of your lower bounds is the better one?

In each part obtain the required minimum spanning tree either by inspection from the diagram (with the given vertex removed), or by using Kruskal's or Prim's algorithm.

Problem 3.11

Consider the travelling salesman problem for the six places in Ireland (see Problems 3.6 and 3.9). Using the above method, find the lower bound obtained by removing the vertex Athlone. (First find a minimum spanning tree for the appropriate set of vertices using Kruskal's or Prim's algorithm or by inspection of a suitable diagram.)

3.5 Computer activities

The computer activities for this section are described in the *Computer Activities Booklet*.

> After studying this section, you should be able to:
> - define the terms *spanning tree, minimum spanning tree, maximum spanning tree, minimum connector* and *maximum connector*;
> - use Kruskal's and Prim's algorithms for solving the minimum connector problem;
> - use Kruskal's and Prim's algorithms for solving the maximum connector problem;
> - obtain a lower bound for the solution to the travelling salesman problem.

4. Multi-terminal flows

In this section we discuss multi-terminal flows in an undirected network, and use a tree algorithm to construct a type of tree called a *cut tree* which enables us to determine the maximum flow between any two vertices of an undirected network.

In *Networks 1* we were primarily concerned with problems involving the flow of a commodity from a single source to a single sink. However, many of the network problems arising in practice involve the flow of commodities between many different pairs of locations. For example, an airline network involves flights between a number of cities; a telephone network links many pairs of subscribers or exchanges; and a road network involves the flow of traffic between a number of different locations. Problems of this type are known as **multi-terminal flow problems**.

Multi-terminal flow problems of the types just described are, in general, too complicated to be solved here. We therefore restrict our attention to one particular type:

> find a maximum flow from any location in a given network to
> any other location in the network, assuming that only a
> single pair of locations can communicate at any one time.

Although this simplified version of the problem may seem too restrictive to be of practical use, it is important in certain applications. For example, the technique we shall describe is very useful in problems involving the expansion or improvement of an existing system such as a road network (widening a road or building an overpass) or a telephone network (constructing a new exchange or increasing the capacity of a particular link).

In order to solve this simplified multi-terminal flow problem, we could proceed by considering separately each possible source and sink, and use the techniques described in *Networks 1* to find a maximum flow in each case. If the number of vertices in the network is n, then the number of ways of choosing the source and sink is $n(n-1)$, since the source can be chosen in n ways, and the sink can then be chosen in $n-1$ ways. It follows that we should have to do $n(n-1)$ separate maximum flow calculations — which is a lot of work!

We can halve the number of calculations by restricting our attention to *undirected* networks, since any flow from a vertex V to a vertex W gives rise to a flow from W to V (just reverse the flow!). So, for an undirected network, we would have to do $n(n-1)/2$ separate maximum flow calculations. But this is still an enormous number — even for networks with only 20 vertices, we would still need to carry out 190 different maximum flow calculations.

Fortunately, we can get away with much less work than this. Provided that we restrict our attention to undirected networks, we can use an algorithm due to R. E. Gomory and T. C. Hu to reduce the number of maximum flow calculations from $n(n-1)/2$ to just $n-1$; for example, a network with 20 vertices would require 19 maximum flow calculations instead of the 190 mentioned above. The approach that Gomory and Hu used is entirely different from the flow algorithms you met earlier, and involves the construction of a particular type of tree, called a *cut tree*.

4.1 Cut trees

We begin by considering a simple example that leads us to formulate a result known as the *capacity rule*.

Example 4.1

Consider the following undirected network N.

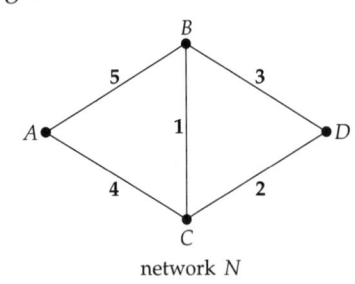

network N

For each pair of vertices V and W of N, we calculate f(VW), the value of a maximum flow between V and W. In order to do this, it is often simplest to look at the various cuts separating V and W, and to find a cut with minimum capacity; by the max-flow min-cut theorem, this capacity gives the required value. For example, in the above network, f(AD), the value of a maximum flow between A and D (or D and A), is 3 + 2 = 5, and f(AB), the value of a maximum flow between A and B (or B and A), is 5 + 1 + 2 = 8.

To define a minimum cut in an *undirected* network, we replace each undirected edge by two arcs of the same capacity as the edge, one in each direction, and use the definition of a minimum cut for a directed network (*Networks 1*, Section 3.1). Thus the capacity of a minimum cut in an undirected network is the sum of the capacities of the edges in the cut (since, for each such edge, *one* of the arcs replacing it will contribute towards the capacity of the cut).

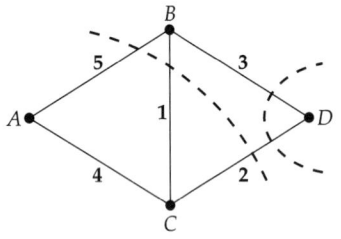

We now construct a new undirected network by taking the *complete* graph with vertices A, B, C and D, and assigning to each edge VW a capacity equal to f(VW); for example, the capacity of the edge AD is 5, and the capacity of the edge AB is 8. We call this new network the **complete network** of N, and denote it by C(N).

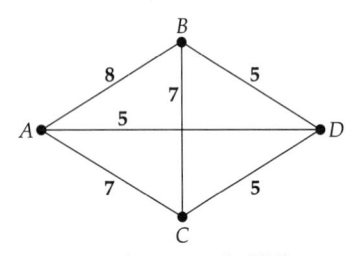

complete network C(N)

There are two features to note about the complete network C(N):

1. in each cycle of C(N), the smallest capacity occurs more than once; for example, in the triangle ABCA the capacity 7 occurs twice, and in the cycle ABDCA the capacity 5 occurs twice;

2. C(N) has four vertices, but only three different capacities (5, 7 and 8).

Problem 4.1

Consider the undirected network N shown in the margin.

(a) Find f(VW) for each pair of vertices V and W.

(b) Construct the complete network C(N), and write down statements analogous to statements 1 and 2 in the above example.

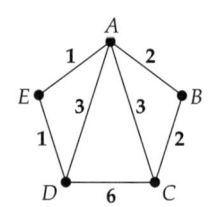

The above statements 1 and 2 are special cases of the following general statements for an undirected network N and its complete network C(N).

1. In each cycle of C(N), the smallest capacity occurs more than once.
2. If N has n vertices, then C(N) has at most n − 1 different capacities.

It follows from Statement 2 that at most $n-1$ of the $n(n-1)/2$ possible values of maximum flows between pairs of vertices are different. Before describing how to calculate these $n-1$ values, we prove Statements 1 and 2.

We first prove Statement 1.

> **Theorem 4.1**
>
> Let N be an undirected network, and let $C(N)$ be the corresponding complete network. Then, in any cycle of $C(N)$, the smallest capacity occurs more than once.

Proof

Let $f(VW)$ be the smallest capacity in some cycle of $C(N)$. Since $f(VW)$ is the value of a maximum flow from V to W in the network N, it follows from the max-flow min-cut theorem that there is a minimum cut in N with capacity $f(VW)$. The removal of the edges in this cut separates the network N into two parts, X (containing V) and Y (containing W).

Now let $V'W'$ be any other edge of the cycle with one end in X and the other end in Y. There must be such an edge, because in going from V to W we cross from X to Y, and we eventually need to return to X in order to complete the cycle — in fact, we may need to cross several times from X to Y in tracing the cycle.

Note that it is possible for V and V', or W and W', to be the same vertex; this happens if, for example, the cycle is a triangle.

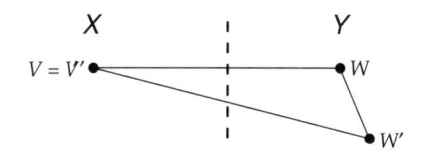

Since the minimum cut separating V from W also separates V' from W', any flow between V' and W' must have value not exceeding $f(VW)$, and so

$f(V'W') \leq f(VW)$.

But $f(VW)$ is the *smallest* capacity occurring in the cycle, so

$f(VW) \leq f(V'W')$.

It follows from these two inequalities that $f(VW) = f(V'W')$, and so the smallest capacity occurs at least twice. ∎

In order to prove Statement 2, we need to construct in $C(N)$ a spanning tree with the property that the sum of the capacities of all its edges is a maximum — such a tree is called a maximum spanning tree, as you saw in Section 3.3. The construction uses Kruskal's algorithm for maximum spanning trees: we successively choose edges of largest capacity in such a way that no cycles are created. The following example illustrates the method.

Alternatively, Prim's algorithm may be used.

Example 4.1 (continued)

Let N and $C(N)$ be as before:

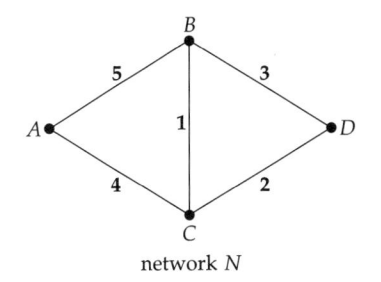

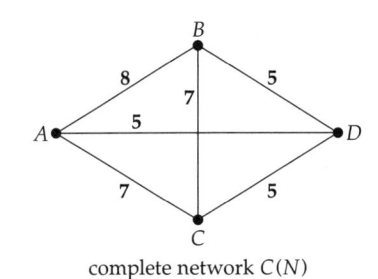

network N

complete network $C(N)$

First edge

We choose an edge of largest capacity in $C(N)$; the only possibility is AB, with capacity $f(AB) = 8$.

Second edge

We choose an edge of next largest capacity; this can be either AC or BC, with capacity $f(AC) = f(BC) = 7$. Let us choose AC.

Third edge

We cannot include both AC and BC in the tree, as this would create a cycle ($ABCA$). So we choose one of the three edges AD, BD or CD, with capacity $f(AD) = f(BD) = f(CD) = 5$. Let us choose AD.

Since the network has four vertices, the required tree has three edges, so this completes the maximum spanning tree.

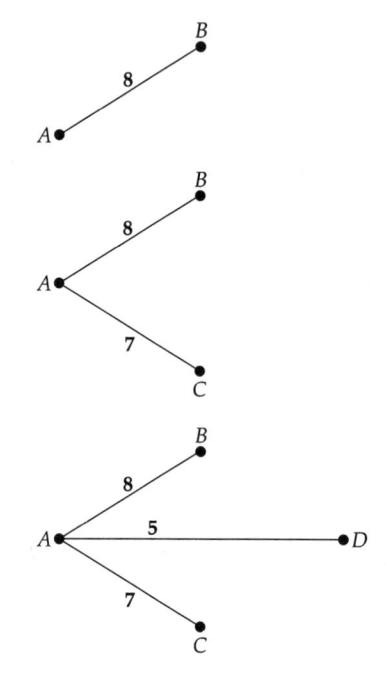

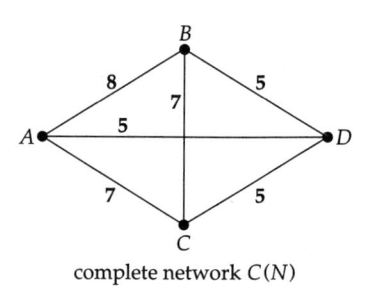

complete network $C(N)$

Note that in this example, there are two possible choices at the second stage, and three possible choices at the third stage. These possibilities give rise to the following six maximum spanning trees, each having three edges with capacities 5, 7 and 8.

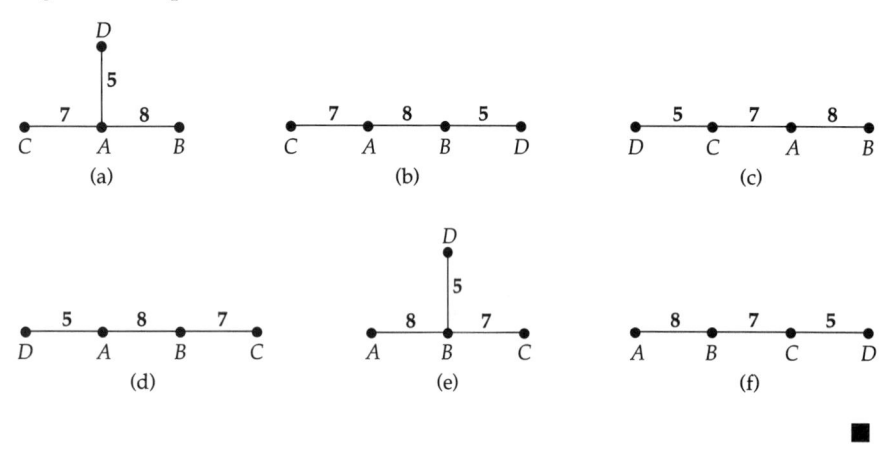

The importance of these maximum spanning trees lies in the fact that we can use them to reconstruct all the capacities in $C(N)$. In order to do this, we use the following rule.

Capacity rule

To find $f(VW)$ for two vertices V and W in an undirected network N, look at the path joining the vertices V and W in a maximum spanning tree of the complete network $C(N)$, and take the smallest of the capacities along this path.

For example, applying the rule to tree (d) above, we obtain

$f(AC) = \min(f(AB), f(BC)) = \min(8, 7) = 7$,

$f(BD) = \min(f(BA), f(AD)) = \min(8, 5) = 5$,

$f(CD) = \min(f(CB), f(BA), f(AD)) = \min(7, 8, 5) = 5$,

which are the correct values.

We write min (x, y) to denote the smaller of the numbers x and y.

Problem 4.2

Find the values of $f(AD)$, $f(BC)$ and $f(CD)$, by applying the capacity rule to tree (b) above.

Problem 4.3

The following diagram shows a maximum spanning tree in a complete network.

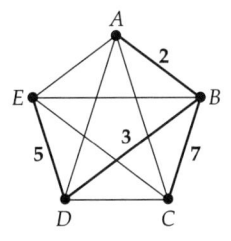

Use the capacity rule to write down the values of $f(AC)$, $f(AD)$, $f(AE)$, $f(BE)$, $f(CD)$ and $f(CE)$.

Problem 4.4

Prove the capacity rule, by applying Theorem 4.1 to the cycle obtained when the edge VW is added to a maximum spanning tree for $C(N)$.

We now use the capacity rule to prove Statement 2.

Theorem 4.2

Let N be an undirected network with n vertices, and let $C(N)$ be the corresponding complete network. Then $C(N)$ has at most $n - 1$ different capacities.

Proof

It follows from the capacity rule that the capacity of any edge of $C(N)$ is equal to the capacity of an edge in a maximum spanning tree for $C(N)$. Since any tree with n vertices has exactly $n - 1$ edges, the capacity of any edge of $C(N)$ must equal the capacity of one of these $n - 1$ edges. The result follows. ∎

As yet, we have not accomplished much as far as the original multi-terminal flow problem is concerned. We have merely shown that, in a network N with n vertices, there are at most $n - 1$ different values of maximum flows $f(VW)$, and that these can easily be found if we know the capacities of the edges in a maximum spanning tree for $C(N)$. However, we cannot consider the original problem solved until we have a method for constructing such a maximum spanning tree directly from the original network.

Shortly, in the audio-tape sequence, we shall present a method for finding such a maximum spanning tree, which constructs a particular type of tree, called a *cut tree*, using an algorithm due to Gomory and Hu. Note that their method works only for *undirected* networks.

The algorithm of Gomory and Hu is based on two main ideas — *flow-equivalent networks* and *cut trees*. The former of these is defined as follows.

Definition

Two undirected networks are **flow-equivalent** if:

(a) they have the same number of vertices;

(b) the maximum flows between pairs of vertices in the two networks are the same.

For a directed network, the maximum flow from a vertex V to a vertex W may be different from the maximum flow from W to V, and no general method is known that can take account of this efficiently in calculating the $n(n - 1)$ different flows.

For example, in Example 4.1 we saw that the network N, repeated here in the margin, is flow-equivalent to each of the following trees.

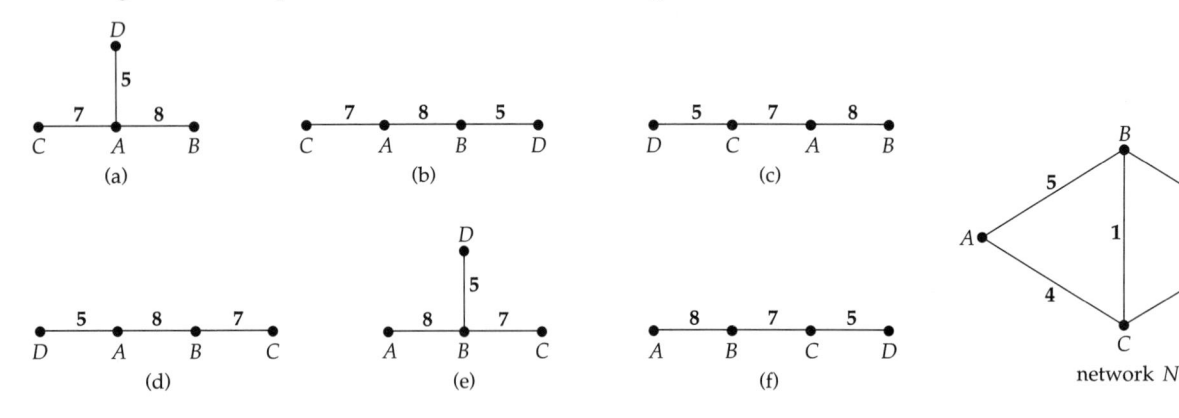

```
     D
     •
     |5
  7  |  8
•————•————•
C    A    B
    (a)

  7     8     5
•————•————•————•
C    A    B    D
     (b)

  5     7     8
•————•————•————•
D    C    A    B
     (c)

  5     8     7
•————•————•————•
D    A    B    C
     (d)

        D
        •
        |5
   8    |   7
•————•————•
A    B    C
    (e)

  8     7     5
•————•————•————•
A    B    C    D
     (f)
```

network N

This example illustrates that for any undirected network N:

- we can find a tree that is flow-equivalent to it by finding a maximum spanning tree for C(N);
- there are several maximum spanning trees for C(N).

The algorithm of Gomory and Hu constructs a particular type of flow-equivalent tree, called a *cut tree*; such trees have a special property involving minimum cuts.

Example 4.1 continued

To see this special property, we look at a minimum cut in our example. Suppose that we find the maximum flow between A and B. The corresponding minimum cut is shown by the dashed line in the diagram below; it separates the vertices of the network into two disjoint sets — X = {A, C} and Y = {B, D}. The capacity of this cut is 5 + 1 + 2 = 8, and so, by the max-flow min-cut theorem, the value of the maximum flow from A to B is also 8.

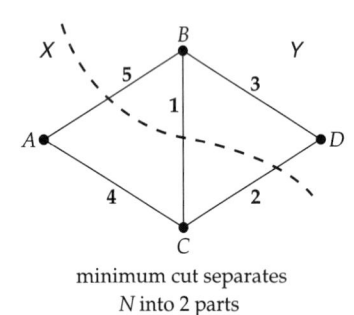

minimum cut separates
N into 2 parts

Now consider the maximum flow from A to B in each of our maximum spanning trees:

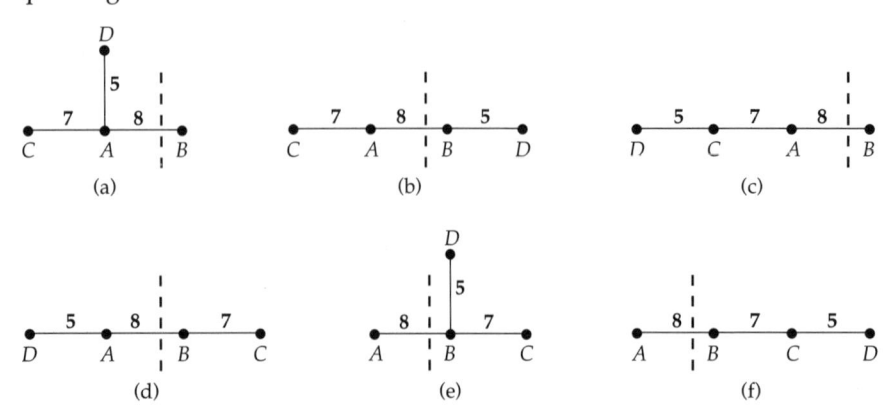

In each case, the corresponding minimum cut is shown by a dashed line through the edge AB. In tree (a), the minimum cut divides the vertices into the two sets {A, C, D} and {B}; in tree (b), the minimum cut divides

44

the vertices into the two sets {A, C} and {B, D}. Note that, for tree (b), the parts of the minimum cut are *the same as in the original network*, but for tree (a), these parts are not the same. If you find the minimum cuts between the other pairs of vertices, you will see that tree (b) has this property for *all* the minimum cuts.

The flow-equivalent tree (b) is rather special. We call it a *cut tree* for the network N. ∎

> **Definition**
>
> A flow-equivalent tree is a **cut tree** if each branch of the tree corresponds to a minimum cut of the original network; that is, each branch, when removed, separates the vertices into two sets X and Y corresponding to a minimum cut in the original network.

Problem 4.5

For the above network N, determine whether any of the above trees (c), (d), (e) and (f) is a cut tree for N.

Problem 4.6

For the above network N and tree (b), find minimum cuts corresponding to:

(a) a maximum flow from C to D;

(b) a maximum flow from B to C.

4.2 Algorithm of Gomory and Hu

In the audio-tape sequence, we present the algorithm of Gomory and Hu for obtaining a *cut tree*.

We first show how the algorithm is used for finding the maximum flows between all pairs of vertices in a given undirected network. The algorithm constructs a cut tree, branch by branch, by calculating a maximum flow for each branch. To attach each new branch to the existing tree, we use the cut tree property: *the minimum cuts in the cut tree and the corresponding minimum cuts in the original network separate the same vertices*. If the original network has n vertices, then our cut tree has $n - 1$ edges, so we need perform only $n - 1$ maximum flow calculations.

We then show how some of these calculations may be simplified by *condensing* certain vertices of the given network into a single vertex.

Once a cut tree has been constructed, the remaining maximum flows may be found by using the capacity rule, since a cut tree is a maximum spanning tree for the corresponding complete network.

> Now listen to band 1 of Audio–tape 2.

> After studying this section, you should be able to:
> - state and use the *capacity rule*;
> - explain the terms *flow-equivalent networks* and *cut tree*;
> - use the algorithm of Gomory and Hu to solve multi-terminal flow problems.

5 Gas pipeline networks

This is the television section. The printed material associated with the programme is given in the *Television Notes Booklet*. You are advised to look at this material before watching the programme.

Appendix: Trees, telephones and tiles

How does a telephone company run cables from town to town along the most economical route? Mathematicians have a simple answer but proving why it's right is more tricky.

Peak District Cable, a new company with big ambitions but a budget more in line with these recessionary times, wishes to lay a network of cables that will link the three towns of Loughborough, Stoke-on-Trent and Rotherham. But its two senior managers, Miles Spanning and Horatio Steiner, cannot agree on how this is to be done. Spanning argues that because straight lines are the shortest paths between points, the best that can be done is to pick one town and link it to each of the others by a straight cable. The choice of town depends upon the actual distances, but it is simple enough to decide: look at the triangle formed by the towns, discard its longest side and lay cable along the other two. Steiner concedes most of this but has got it into his head that adding an extra town to the network might actually make it shorter. It may seem unlikely that he is right—surely extra towns need extra cable—but if he is, where should the extra town be located to save the most cable, and how much does it save?

As it happens, Rotherham, Loughborough and Stoke-on-Trent are all the same distance apart, 75 kilometres. Spanning's network will be 150 kilometres long, no matter which of the three towns is used as the central link. But what about Steiner's harebrained suggestion? Nestling in the Derbyshire dales near Matlock, roughly in the middle of the triangle formed by the three towns, is a village called Middleton. It is roughly 44 kilometres from each town. Steiner points out that if the centre of the network is located at Middleton, and one link is run from there to each of the three towns, then the total will be 3 × 44 = 132 kilometres, a saving of 18 kilometres, or about 12 per cent (see Figure 1).

The same approach can be used with any collection of towns. You can use Spanning's method, and find the shortest network that joins them with straight links without introducing extra towns; or you can follow Steiner and "invent" new towns, which may shorten the network if you put them in the right place. In 1968 Edgar Gilbert and Henry Pollak, of AT&T Bell Laboratories at Murray Hill, New Jersey, conjectured that no matter how the towns are initially chosen, the maximum saving in cable that can be obtained by adding extra towns is 13.34 per cent. Following the usual rule for mathematical attribution, which is to name the idea after someone vaguely connected with the problem, this has become known as the "Steiner ratio conjecture".

After 23 years of unrelenting effort, Ding Zhu Du at Princeton University and Frank Hwang at Bell Labs have proved the conjecture to be true. In its mathematical formulation the towns to be connected are represented by points in a plane, and the cables linking them are straight lines. Whether or not we invent new towns, it is clear that the links must form a tree—a network without any loops. Loops just waste cable, joining up towns that are already joined by some other route.

If no new towns are involved, then such a network is called a "spanning tree". There are a lot of spanning trees to choose from, but in principle you can just list them all and see which is the shortest. For example, suppose there are four towns: Aylesbury, Brighton, Clacton, and Dagenham.

This article by **Ian Stewart** appeared in *New Scientist*, 16 November 1991, pp. 26–9. Ian Stewart is Professor of Mathematics at the University of Warwick.

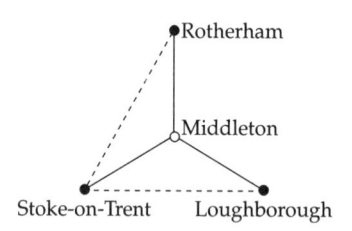

Figure 1 How to save cable by adding a town. The dashed network is 150 kilometres long, the solid one is only 132 kilometres.

Figure 2 shows some of the possible spanning trees and their lengths. The shortest one has a "branch point" at Dagenham, from which three links run to the other three towns. On the other hand, if the towns are Ashburton, Bristol, Cheltenham and Daventry, which are located roughly along a straight line, then you can easily convince yourself that the shortest spanning tree joins them in that order and has no branch points.

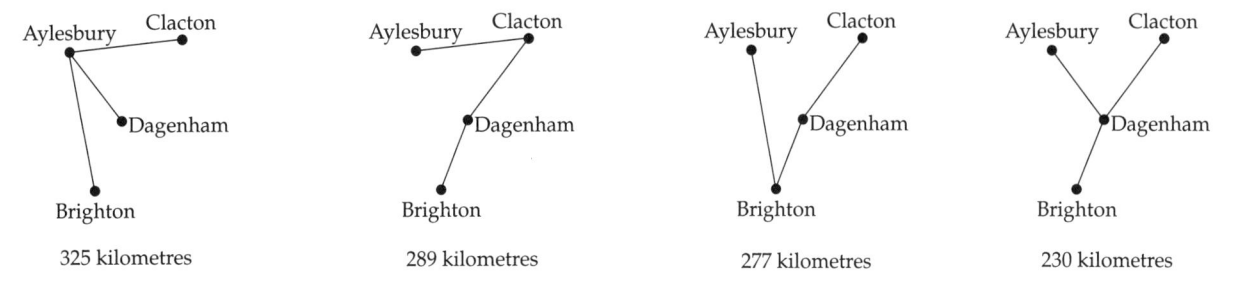

325 kilometres   289 kilometres   277 kilometres   230 kilometres

Figure 2
Four of the 16 possible spanning trees for four English towns. The one on the right is the shortest among all 16.

The problem is far more subtle if links are allowed to meet out in the countryside as well. For example, if there are three towns at the vertices of an equilateral triangle, as in our opening example of Rotherham, Loughborough and Stoke-on-Trent, then the shortest network joins all three to the centre of the triangle at Middleton—or, in reality, a nondescript place somewhere in a field not far from it. The shortest such network must also be a tree: it is called a "Steiner tree". In general a Steiner tree is a tree whose edges either meet at the original towns, or at new towns, in which case the angles between the lines must be 120°. The shortest tree that joins the original towns, including possible new ones, is always a Steiner tree, so the search for the shortest tree can be confined to Steiner trees. However, there are in general many Steiner trees, some longer than others.

Who was the real Steiner, and what does he have to do with the problem? As it happens, not much. Jakob Steiner was a 19th-century Swiss mathematician, who in 1837 solved the problem linking three towns. His name was attached to the problem by Richard Courant and Herbert Robbins in 1941, in their classic popularisation, *What is Mathematics?* But neither they, nor Steiner himself, seem to have known that he had been beaten to the punch by Evangelista Torricelli and Francesco Cavalieri in 1640 or thereabouts, when they broke the problem down into two different cases. If the triangle has an angle of 120° or more, then the shortest network consists of just two links, joining that vertex to the other two. But if the angles of the triangle formed by the three towns are all less than 120°, then the shortest network consists of three links that lead from the towns to the "Steiner point"—the unique place where all three roads meet at angles of 120° (see Figure 3). Later Steiner also proved that when there are several towns, the edges of any Steiner tree must meet at 120° at each new town added, a simple consequence of the solution for three towns. He was less forthcoming on how to find such trees.

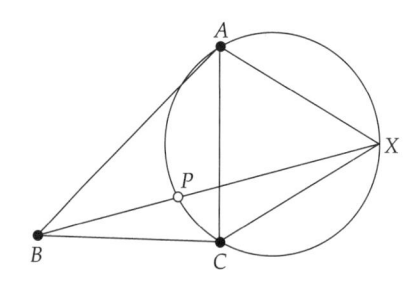

Figure 3 Constructing the Steiner point of triangle *ABC*. Draw an equilateral triangle *ACX*. Its circumcircle cuts *BX* at the Steiner point *P*.

The problem of using a Steiner tree to join more than three towns was first investigated seriously by Milos Kossler and Vojtech Jarnik in 1934. Finding the shortest Steiner tree in any given example requires a much more complicated calculation than finding the shortest spanning tree, because so many new Steiner points have to be considered. For example, suppose there are six towns arranged at the corners of two adjacent squares, as in Figure 4. One possible Steiner tree is shown in Figure 4(a): it is found by solving the problem for a square of four towns first, and then linking in the two remaining towns via their Steiner point with one that is already linked. The shortest Steiner tree, however, is that shown in Figure 4(b). You cannot build up the shortest possible Steiner tree piecemeal.

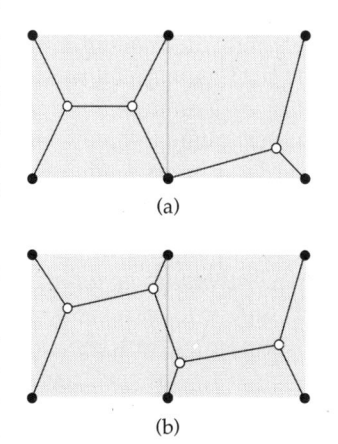

Figure 4 Combining Steiner trees for a square and an isosceles right triangle gives (a), but (b) gives a shorter Steiner tree for the same set of towns.

Gilbert and Pollak asked whether the two versions of the problem might be related. Call the length of the shortest spanning tree the *spanning length* of the set of towns, and that of the shortest Steiner tree the *Steiner length*. Now, every spanning tree is also a Steiner tree (invent no new towns at all). So, for any set of towns, the spanning length is always greater than or equal to the Steiner length. How much greater?

For an equilateral triangle of unit side, the spanning length is 2 and the Steiner length is $\sqrt{3}$. In this case, the ratio between the Steiner length and the spanning length is $\sqrt{3}/2$, or 0.8666, and the saving in length obtained by using the shortest Steiner tree rather than the shortest spanning tree is about 13.34 per cent. Gilbert and Pollak's Steiner ratio conjecture states that you can never do better than this. For any number of towns, arranged in any possible manner, the ratio of the Steiner length to the spanning length is always greater than or equal to $\sqrt{3}/2$. Problems that are similar in principle, though more complicated in detail, arise in the routeing of telephone lines, gas pipes, cable television networks and public transport—and also in one approach to the evolution of living organisms.

The genetic material of living creatures is DNA, which "encodes" developmental information as a sequence of the four bases adenine, thymine, cytosine and guanine. In this scheme, genetic information is specified by a long sequence of bases, such as AATTCGCTCA... In the application of Steiner trees, the "towns" are sequences of DNA in different organisms, and the "distance" is some measure of the similarity between different sequences, such as the proportion of corresponding bases that are equal. Steiner points correspond to "most plausible common ancestors". Although there is no guarantee that these common ancestors existed, the method provides interesting clues to how the DNA molecule might have changed and how organisms are related genetically.

Shorter circuits get on board

Some of the other problems analogous to Steiner trees are practical in nature. In the design of electronic circuits, for example, the connections are generally laid out on a rectangular grid, running only horizontally or vertically. Here the same kinds of questions can be asked as with other applications, and similar methods may help with their solutions.

The Steiner ratio conjecture is important for the "economics" of all such networks, because the shortest spanning tree is much easier to find than the shortest Steiner tree, and it may therefore be worth sacrificing the 13.34 per cent error to save computational effort. It is no coincidence that AT&T, which owns the labs from which the Steiner ratio conjecture has emerged, is a telephone company. Until quite recently, AT&T used the spanning length as a convenient way to charge customers who wanted to connect their offices together. It was worried that customers might discover that they could make major savings on their bills by inventing imaginary offices located at appropriate Steiner points. The conjecture limits such savings to 13.34 per cent, which is not too embarrassing for the company. In principle, AT&T could have saved itself the worry and used the Steiner length instead. But it couldn't.

Finding the spanning length is a simple computation, even for a huge number of towns. It is solved by the "greedy" algorithm: start with the shortest link you can find, and at each stage thereafter add on the shortest remaining link that does not complete a closed loop, until every town is included in the tree. (An algorithm is a specific computational procedure that is guaranteed to give an answer.)

Finding the shortest Steiner tree is not so easy. You cannot do it just by taking all possible triples of towns, finding their Steiner points, and then looking for the shortest tree that joins the towns together and meets either at towns or at Steiner points. For example, consider four towns (black) in a square (see Figure 5). The Steiner points for triples of towns are tucked in near the corners (the grey blob in the figure and three corresponding points near the other corners). But there is another Steiner tree, using two new towns indicated by the white points. The white points are not Steiner points of any three of the towns. Moreover, the tree that uses the white points can be proved to be the shortest tree. The meaning of the phrase "Steiner point" for more than three towns is quite complicated.

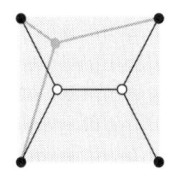

Figure 5 Steiner points (white) for four towns (black) in a square are different from the Steiner points (grey) of a subset of three towns.

Most points are not Steiner points in this more complicated sense, but it is not easy to decide which are which. There are infinitely many points in the plane, so a finite algorithm cannot check them one by one. This makes it far from obvious that finite algorithms exist at all. In fact such algorithms do exist, and the first was invented by Z. A. Melzak of the University of British Columbia. But his method becomes unwieldy even for moderate numbers of towns, and improved versions devised since are not much better.

We now know that there are good reasons why these solutions are inefficient. The growing use of computers has led to the development of a new branch of mathematics called complexity theory, which studies not just algorithms, but how efficient those algorithms are. Given a problem involving some number n of objects (towns in this case), how fast does the running time of the solution grow as n grows? If the running time grows no faster than a constant multiple of a fixed power of n, such as $5n^2$ or $1066n^4$, then the algorithm is said to run in polynomial time, and the problem is considered to be "easy". Usually this means that the algorithm is practical though it will not be if the constant is absolutely huge. If the running time grows faster than any constant multiple of powers of n—for instance exponentially, like 2^n or 10^n—then the problem is said to have a nonpolynomial running time and is "hard". Usually this means that the algorithm is totally impractical. In between polynomial time and exponential time is a wilderness of "fairly easy" or "moderately hard" problems, where practicality is more a matter of experience.

For instance, adding two n-digit numbers requires at most $2n$ one-digit additions, including carries, so the time taken is "bounded" by (is never greater than) a constant multiple (namely 2) of the first power of n. Long multiplication of two such numbers involves about n^2 one-digit multiplications and no more than $2n^2$ additions, or $3n^2$ operations on digits, so now the bound involves only the second power of n. These problems are therefore "easy", whatever generations of schoolchildren may think. In contrast, consider the travelling salesman problem, which is to find the shortest route that will take a salesman through a given set of cities. If there are n cities, then the number of routes that we have to consider is $n!$, or $= n \times (n-1) \times (n-2) \times \cdots \times 3 \times 2 \times 1$, which grows faster than any power of n. Case-by-case enumeration is therefore hopelessly inefficient.

Oddly enough, the big problem in complexity theory is how to prove that the subject actually exists. That is, to prove that some apparently interesting problem really is hard. The difficulty is that it is easy to prove a problem is easy, but hard to prove that it is hard. To show a problem is easy, you just exhibit one algorithm that solves it in polynomial time. It does not have to be the best, or the cleverest: any will do. But to prove that a problem is hard, it is not enough to exhibit an algorithm with nonpolynomial running time. You may have chosen a silly one, or there may be a better one that does run in polynomial time. You have to find some mathematical way to consider all possible algorithms for the problem, and show that none of them runs in polynomial time.

There are lots of candidates for hard problems, including the travelling salesman problem. There is also the bin-packing problem of how to best fit a set of items of given sizes into a set of sacks of given sizes. And there is the knapsack problem: given a fixed-size sack and lots of objects, does any set of objects fill the bag exactly? So far, nobody has managed to prove any of them are hard. However, in 1971 Stephen Cook of the University of

Toronto showed that if you can prove that any one problem in this candidate group really is hard, then they all are. Roughly speaking, you can "code" any one of them to become a special case of one of the others: they sink or swim together. These problems are called "NP-complete", where NP stands for nonpolynomial. Everyone believes they really are hard, if only on the not very mathematical grounds that you cannot expect to get all good things at once.

Ron Graham, Michael Garey and David Johnson of AT&T have proved that the problem of the Steiner length is NP-complete. An efficient algorithm to find the precise Steiner length for any set of towns would automatically lead to efficient solutions to all sorts of computational problems that are widely believed not to possess such solutions. The Steiner ratio conjecture is therefore important, because it proves that you can replace a hard problem by an easy one without simplifying it very much. The equilateral triangle example that started the whole business is very natural. It suggests that there must be a simple proof of the conjecture. However, its simplicity may be deceptive, for if there is a simple proof of the Steiner ratio conjecture, nobody has ever found it. Even the Du-Hwang proof is quite tricky. And, leaving their proof aside, direct attacks for small numbers of towns lead to vast and messy calculations. Gilbert and Pollak had quite a lot of evidence for their conjecture. In particular, they could show that something along those lines must be true: they proved that the ratio is always at least 0.5. By 1990, various people had performed heroic calculations to verify the conjecture completely for networks of four, five and six towns. For general arrangements of as many towns as you like, they also pushed up the limits on the ratio from 0.5 to 0.57, 0.74, and 0.8. Not long ago Graham and Fan Chung at Bell Communications Research raised it to 0.824, in a computation that they describe as "really horrible—it was clear it was the wrong approach".

To make further progress possible, the horrible calculations had to be simplified. Du and Hwang found an approach that does away with the horrible calculations completely. The basic question is how to get equilateral triangles in on the act. There is a big gap between the triangle example, which sets up the bound on the ratio, and a general system of towns, which is supposed to obey the same bound. How can this no-man's-land be crossed? There is a kind of halfway house. Imagine the plane tiled with identical equilateral triangles, in a triangular lattice pattern (see Figure 6). Put towns only at the corners of the tiles. It turns out that the only Steiner points that need be considered are located in the centres of the tiles. In short, you have a lot of control, not just on computations, but on theoretical analyses.

Of course, not every set of towns lies conveniently on a triangular lattice. Du and Hwang's insight is that the crucial ones do. Suppose the conjecture is false. Then there must exist a counterexample: some set of towns for which the ratio is less than $\sqrt{3}/2$. They show that if a counterexample to the conjecture exists, then there must be one for which all the towns lie on a triangular lattice. This introduces an element of regularity into the problem, and it is then relatively simple to polish it off.

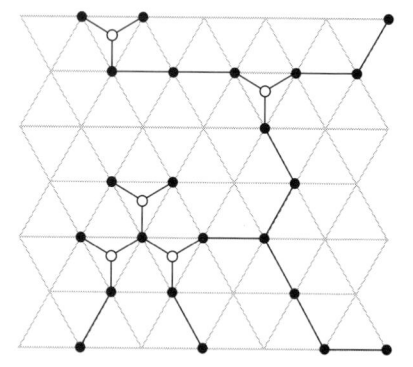

Figure 6 A Steiner tree for towns that lie on a triangular lattice has a much more rigid and regular structure than that for general towns. Du and Hwang reduced the Steiner ratio conjecture to the same problem for lattice trees.

How to prove this lattice property? It is a wonderful exercise in lateral thinking. First, they reformulate the conjecture as a "minimax" problem. Such problems arise in game theory, where players compete and try to limit (minimise the maximum of) the gains (payoff) made by their opponents. Game theory was invented by John von Neumann and Oskar Morgenstern in their classic paper of 1947, *Theory of Games and Economic Behaviour*. In the Du-Hwang version of the Steiner ratio conjecture, one player selects the general "shape" of the Steiner tree, and the other chooses the shortest one of that shape that they can find. Du and Hwang deduce the existence of a lattice counterexample by observing that the payoff for their game has a special "convexity" property.

This elegant new method neatly disposes of a question that previously looked totally intractable, and cuts through a mass of tangled calculations and case-by-case investigations to give a "clean", conceptually simple solution. The Du-Hwang method is not easy, and it does require a certain amount of mathematical technique; but it is such a dramatic improvement that it knocks all previous approaches on the head. More importantly, it provides a paradigm for investigating analogous questions. Formulate the problem in game-theory terms, prove a suitable convexity property, reduce it to a combinatorially "rigid" question with many fewer possibilities—and then solve that, by whatever method you can think of. The motto: "Think first, calculate later": should be engraved on the heart of every mathematician.

Further reading

The material in Section 1.3 is discussed in the article:

C. Alexander, 'A city is not a tree', *Design* **206**, 1966, pp. 44–55.

The material in Sections 1.1, 2 and 3 is covered in most books on graph theory — for example, those listed at the end of *Graphs 1*.

Tree-counting techniques are given in the following.

F. Harary and E. M. Palmer, *Graphical Enumeration*, Academic Press, 1973.

J. W. Moon, *Counting Labelled Trees*, Canadian Mathematical Monographs **1**, Canadian Mathematical Congress, 1970.

G. Pólya and R. C. Read, *Combinatorial Enumeration of Groups, Graphs and Chemical Compounds*, Springer-Verlag, 1987.

F. Harary, 'The number of linear, directed, rooted and connected graphs', *Transactions of the American Mathematical Society* **78**, 1955, pp. 445–63.

A historical discussion of Cayley's work on trees, of the use of trees in chemistry, and of Prüfer's proof of Cayley's theorem, is given in:

N. L. Biggs, E. K. Lloyd and R. J. Wilson, *Graph Theory 1736–1936*, Clarendon Press, 1976, 1986 (pbk).

The minimum connector problem and the travelling salesman problem are discussed in the following.

G. Chartrand and O. R. Oellermann, *Applied and Algorithmic Graph Theory*, McGraw-Hill, 1993.

A. Tucker, *Applied Combinatorics* (2nd edn), Wiley, 1984.

N. Christofides, *Graph Theory — An Algorithmic Approach*, Academic Press, 1975.

Acknowledgements

p.14 text and diagrams from C. Alexander, 'A city is not a tree', *Design* **206**, 1966, pp. 44–55, © 1966 Architectural Forum. Used with permission from Architectural Forum.

p.25 line drawing of Arthur Cayley, courtesy of *Illustrated London News*.

p.25 poem by B. Descartes, courtesy of Professor W. T. Tutte, Waterloo, Ontario, Canada.

p.46 I. Stewart, 'Trees, telephones and tiles', *New Scientist* **132**, 1795, 16 November 1991, IPC Magazines Ltd.

Exercises

Section 1

1.1 A **forest** is a graph (not necessarily connected), each of whose components is a tree.

(a) If G is a forest with n vertices and k components, how many edges does G have?

(b) Construct a forest with twelve vertices and nine edges.

(c) Is it true that every forest with k components has at least $2k$ vertices of degree 1?

Section 2

2.1 Draw as many as you can of the twenty-three unlabelled trees with eight vertices.

The unlabelled trees with seven vertices are given in Solution 1.2.

2.2 Show that there are 125 labelled trees with five vertices.

2.3

(a) Write down the Prüfer sequence associated with each of the following labelled trees.

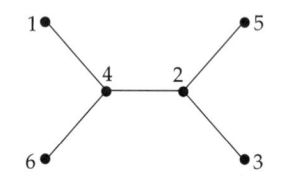

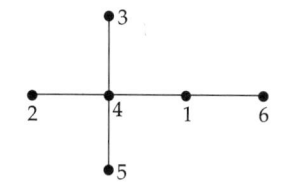

(b) Draw the labelled tree associated with each of the Prüfer sequences $(1, 2, 3, 4)$ and $(3, 3, 3, 3)$.

2.4 Classify all the trees with seven vertices as *central* or *bicentral*.

See Solution 1.2 for these trees.

2.5

(a) There are two different molecules with chemical formula C_3H_7OH. Draw the graphs representing these molecules, and verify that each is a tree.

A carbon atom has valency 4, an oxygen atom has valency 2, and a hydrogen atom has valency 1.

(b) Prove that the graph of any alcohol $C_nH_{2n+1}OH$ is necessarily a tree.

Section 3

3.1 Find all the spanning trees in each of the following graphs.

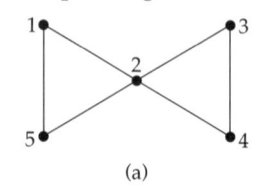

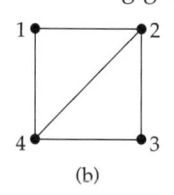

3.2 How many spanning trees has $K_{2,3}$? $K_{2,100}$?

3.3 A **spanning forest** in a graph G (not necessarily connected) is obtained by constructing a spanning tree for each component of G.

(a) Find a spanning forest for the following graph.

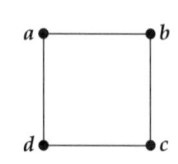

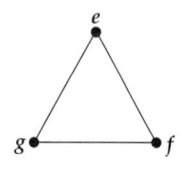

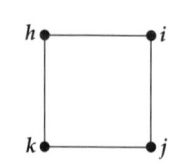

52

(b) Let *G* be a graph, and let *F* be a subgraph of *G*. If *F* is a forest which includes all vertices of *G*, is *F* necessarily a spanning forest of *G*?

3.4 The following table gives the distances (in miles) between five towns.

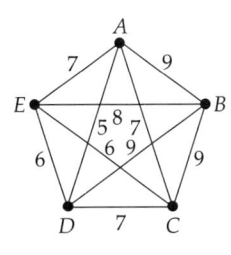

	A	B	C	D	E
A	–	9	7	5	7
B	9	–	9	9	8
C	7	9	–	7	6
D	5	9	7	–	6
E	7	8	6	6	–

(a) Find a minimum connector for these towns, using:

  (1) Kruskal's algorithm;

  (2) Prim's algorithm, starting at vertex *A*;

  (3) the tabular (matrix) form of Prim's algorithm, starting at vertex *A*.

(b) By removing

  (1) town *B*,   (2) town *E*,

  find two lower bounds for the solution to the travelling salesman problem for these towns.

  Find the correct solution by inspection.

3.5 Use the method described in the text to find two lower bounds for the solution to the travelling salesman problem for the following cities.

This data was used in Examples 3.2–3.5.

	Berlin	London	Moscow	Paris	Rome	Seville
Berlin	–	7	11	7	10	15
London	7	–	18	3	12	11
Moscow	11	18	–	18	20	27
Paris	7	3	18	–	9	8
Rome	10	12	20	9	–	13
Seville	15	11	27	8	13	–

Which of your lower bounds is the better one?

Section 4

4.1 Consider the following undirected network *N*.

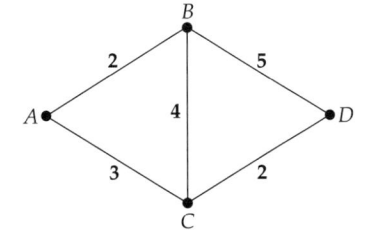

(a) Construct the corresponding complete network *C(N)*, and verify that

  (1) in each cycle of *C(N)*, the smallest capacity occurs more than once;

  (2) *C(N)* has only three different capacities.

(b) Construct a maximum spanning tree for *C(N)*.

4.2 In a multi-terminal flow problem, the following maximum spanning tree was obtained.

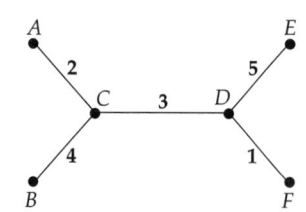

Use the capacity rule to find the value of a maximum flow between each pair of vertices.

4.3 Use the algorithm of Gomory and Hu *without condensing vertices* to find a cut tree for the following undirected network, and hence determine the value of a maximum flow between each pair of vertices. Start by finding a maximum flow from *A* to *E*.

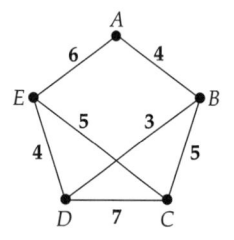

4.4 Consider the following undirected network.

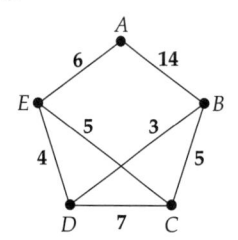

What networks do we get if we condense

(a) the vertices *C, D* and *E*?

(b) the vertices *A* and *B*?

4.5 Use the algorithm of Gomory and Hu to find a cut tree for the network in Exercise 4.4. Start by finding a maximum flow from *A* to *E*. *Hint* Use the condensed networks found in the solution to Exercise 4.4.

54

Solutions to the exercises

1.1

(a) For each component of G, the number of vertices exceeds the number of edges by 1. Since there are k components, the total number of edges is $n - k$.

(b) By the result of part (a), any forest with twelve vertices and nine edges has exactly three components. An example of such a forest is:

(c) If each component has at least two vertices, then the statement is true, as can be seen by applying the result of Problem 1.4(b) to each component. However, the statement is not true in general; for example, the following forest has three components, but only two vertices of degree 1.

2.1 The twenty-three unlabelled trees with eight vertices are:

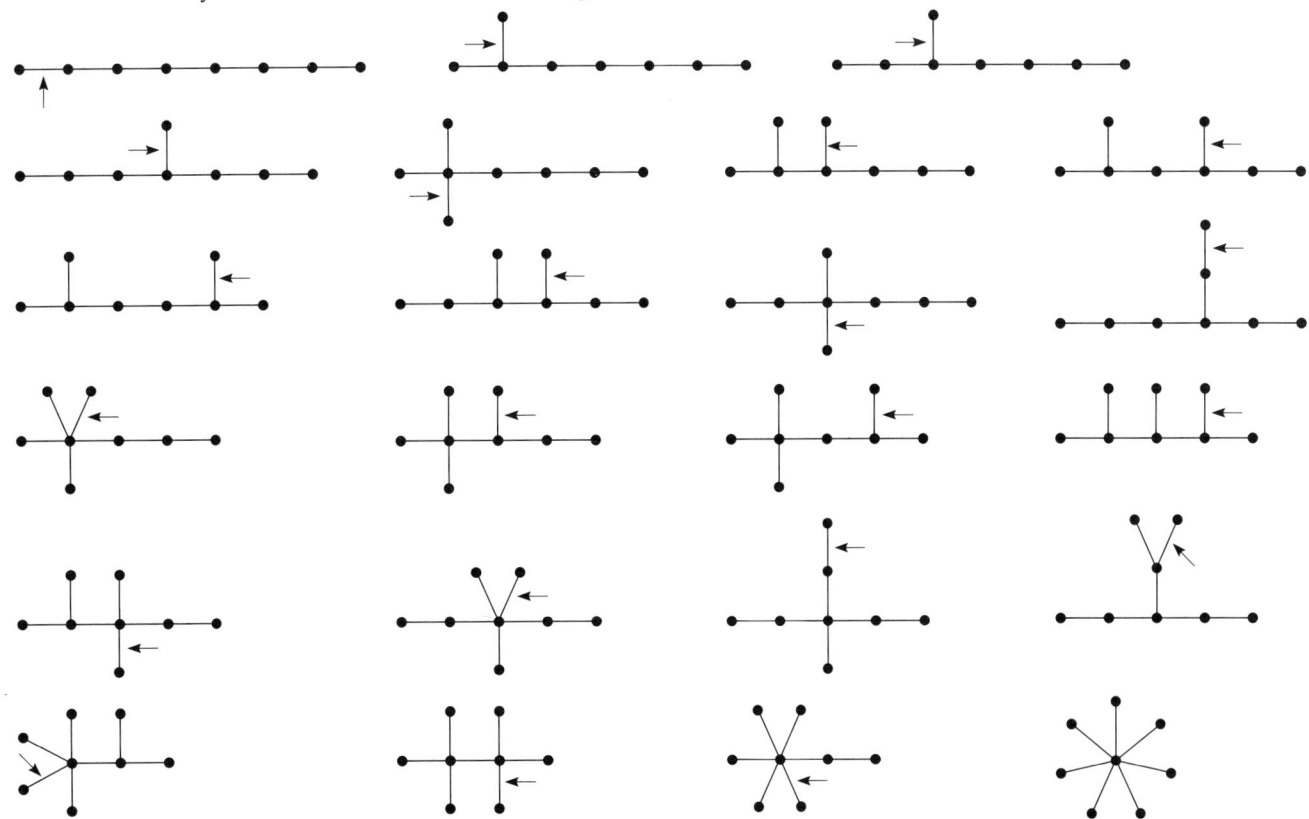

2.2 There are three different *unlabelled* trees with five vertices:

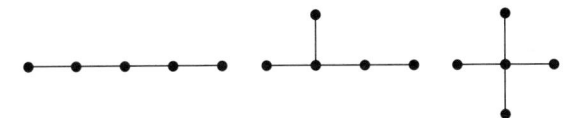

The first of these can be labelled in $\frac{1}{2}(5!) = 60$ different ways, the second can be labelled in $5 \times 4 \times 3 = 60$ different ways, and the third can be labelled in 5 different ways. This gives a total of 125 labelled trees.

55

2.3

(a) (4, 2, 2, 4); (4, 4, 4, 1).

(b)

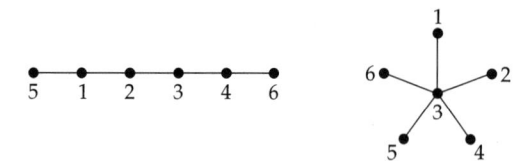

2.4 All trees with seven vertices are central, except for the following four.

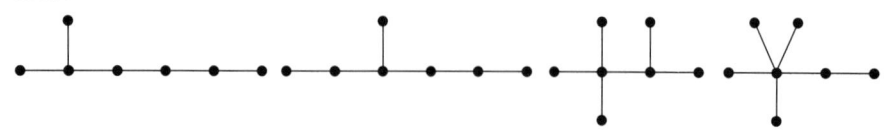

2.5

(a) The structures of these molecules are as follows. Note that in one structure the OH radical is joined to the middle carbon atom, and in the other structure it is joined to an end carbon atom.

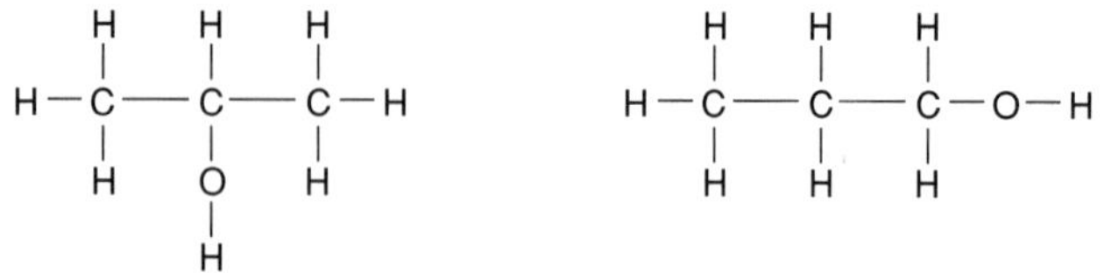

The corresponding graphs are both trees:

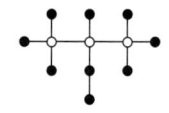

(b) The total number of vertices in the graph of any alcohol $C_nH_{2n+1}OH$ is $n + (2n + 1) + 1 + 1 = 3n + 3$.

Since the oxygen vertex has degree 2, the total number of edges is, by the handshaking lemma,

$$\tfrac{1}{2}[(n \times 4) + ((2n + 1) \times 1) + (1 \times 2) + (1 \times 1)]$$
$$= \tfrac{1}{2}[4n + (2n + 1) + 2 + 1]$$
$$= 3n + 2.$$

Since the graph is connected, and the number of vertices exceeds the number of edges by 1, the graph is a tree, by Theorem 1.1.

3.1 There are nine spanning trees in graph (a):

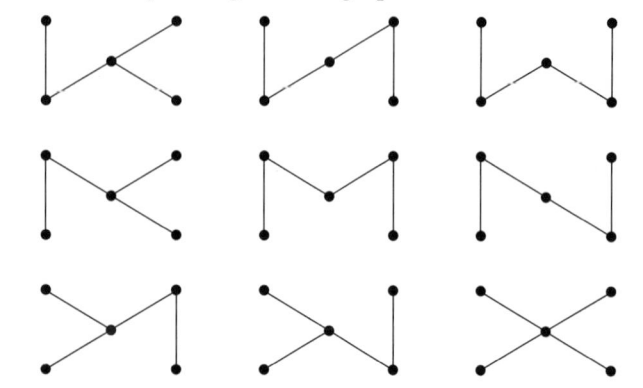

There are eight spanning trees in graph (b):

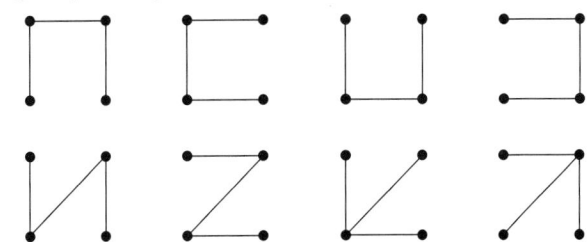

3.2 In $K_{2,3}$ there are twelve spanning trees:

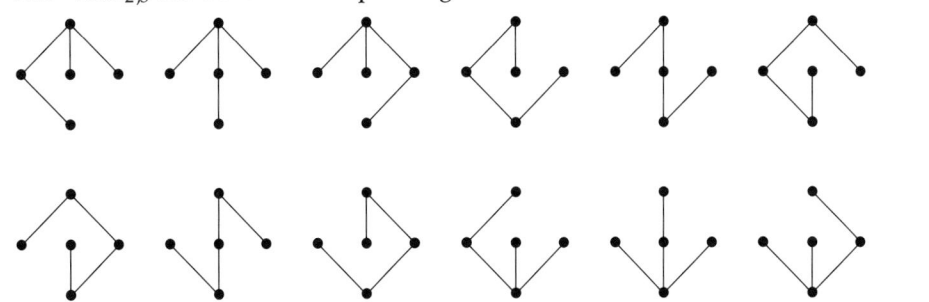

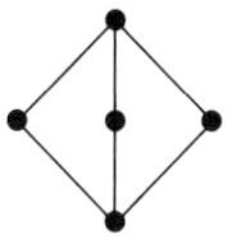

$K_{2,3}$

In $K_{2,100}$ any spanning tree has the form indicated by thick edges in the following diagram.

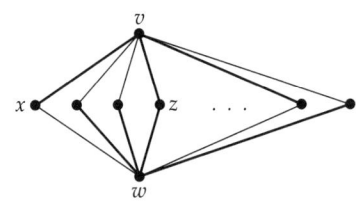

The vertex z joined to both v and w can be chosen in 100 ways, and each of the other 99 central vertices (such as x) can be joined to either v or w — a choice of two possibilities. So the number of spanning trees is 100×2^{99}.

3.3

(a) There are several possibilities — for example:

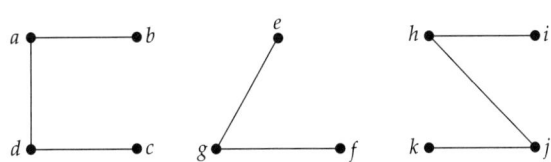

(b) No; for example, the following forest includes all vertices of the graph in part (a), but is not a spanning forest.

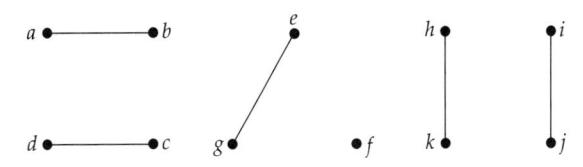

3.4

(a) (1) Using Kruskal's algorithm, we choose edges as follows.

First edge   We choose *AD* (weight 5).

Second edge   We choose *CE* or *DE* (weight 6); let us choose *CE*.

Third edge   We choose *DE* (weight 6); if we had chosen *DE* for the second edge, then *CE* would have been the third choice.

Fourth edge   We cannot choose *AC*, *AE* or *CD* (weight 7), as these would create cycles, so we choose *BE* (weight 8).

This gives the minimum connector of weight $5 + 6 + 6 + 8 = 25$, shown in the margin.

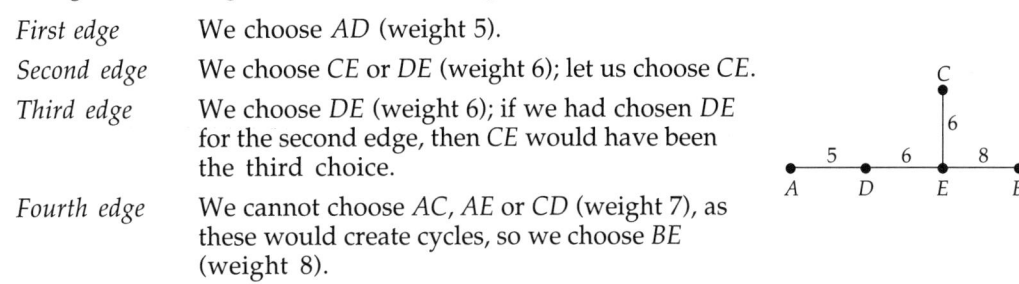

(2) Using Prim's algorithm starting at vertex A, we choose edges as follows.

First edge   We choose AD (weight 5).

Second edge   This time, we have to choose DE (weight 6), as we must join A or D to another vertex.

Third edge   We choose CE (weight 6).

Fourth edge   We choose BE (weight 8).

This gives the same minimum connector as in part (1).

(3) We start with the matrix:

$$\begin{array}{c} & \begin{array}{ccccc} 1 & 2 & 3 & 4 & 5 \end{array} \\ \begin{array}{c} 1 \\ 2 \\ 3 \\ 4 \\ 5 \end{array} & \left[\begin{array}{ccccc} - & 9 & 7 & 5 & 7 \\ 9 & - & 9 & 9 & 8 \\ 7 & 9 & - & 7 & 6 \\ 5 & 9 & 7 & - & 6 \\ 7 & 8 & 6 & 6 & - \end{array} \right] \end{array}$$

STEP 1 We delete all entries in column 1, and mark row 1 with a star.

$$\begin{array}{c} & \begin{array}{ccccc} 1 & 2 & 3 & 4 & 5 \end{array} \\ \begin{array}{c} 1 \\ 2 \\ 3 \\ 4 \\ 5 \end{array} & \left[\begin{array}{ccccc} - & 9 & 7 & 5 & 7 \\ - & - & 9 & 9 & 8 \\ - & 9 & - & 7 & 6 \\ - & 9 & 7 & - & 6 \\ - & 8 & 6 & 6 & - \end{array} \right] \end{array} \begin{array}{c} * \\ \\ \\ \\ \end{array}$$

First iteration

STEP 2 The smallest entry in the starred row is 5 in column 4, so we select 5.

STEP 3 We circle the 5 in row 1 and column 4, mark row 4 with a star, and delete the remaining entries in column 4.

$$\begin{array}{c} & \begin{array}{ccccc} 1 & 2 & 3 & 4 & 5 \end{array} \\ \begin{array}{c} 1 \\ 2 \\ 3 \\ 4 \\ 5 \end{array} & \left[\begin{array}{ccccc} - & 9 & 7 & ⑤ & 7 \\ - & - & 9 & - & 8 \\ - & 9 & - & - & 6 \\ - & 9 & 7 & - & 6 \\ - & 8 & 6 & - & - \end{array} \right] \end{array} \begin{array}{c} * \\ \\ \\ * \\ \end{array}$$

Second iteration

STEP 2 The smallest uncircled entry in the starred rows is 6 in row 4 and column 5, so we select 6.

STEP 3 We circle 6 in row 4 and column 5, mark row 5 with a star, and delete the remaining entries in column 5.

$$\begin{array}{c} & \begin{array}{ccccc} 1 & 2 & 3 & 4 & 5 \end{array} \\ \begin{array}{c} 1 \\ 2 \\ 3 \\ 4 \\ 5 \end{array} & \left[\begin{array}{ccccc} - & 9 & 7 & ⑤ & - \\ - & - & 9 & - & - \\ - & 9 & - & - & - \\ - & 9 & 7 & - & ⑥ \\ - & 8 & 6 & - & - \end{array} \right] \end{array} \begin{array}{c} * \\ \\ \\ * \\ * \end{array}$$

Third iteration

STEP 2 The smallest uncircled entry in the starred rows is 6 in row 5 and column 3, so we select 6.

STEP 3 We circle 6 in row 5 and column 3, mark row 3 with a star, and delete the remaining entries in column 3.

$$\begin{array}{c} & \begin{array}{ccccc} 1 & 2 & 3 & 4 & 5 \end{array} \\ \begin{array}{c} 1 \\ 2 \\ 3 \\ 4 \\ 5 \end{array} & \left[\begin{array}{ccccc} - & 9 & - & ⑤ & - \\ - & - & - & - & - \\ - & 9 & - & - & - \\ - & 9 & - & - & ⑥ \\ - & 8 & ⑥ & - & - \end{array} \right] \end{array} \begin{array}{c} * \\ \\ * \\ * \\ * \end{array}$$

Fourth iteration

STEP 2 The smallest uncircled entry in the starred rows is 8 in row 5 and column 2, so we select 8.

STEP 3 We circle 8 in row 5 and column 2, mark row 2 with a star, and delete the remaining entries in column 2.

$$\begin{array}{c} \\ 1 \\ 2 \\ 3 \\ 4 \\ 5 \end{array} \begin{bmatrix} 1 & 2 & 3 & 4 & 5 \\ - & - & - & ⑤ & - \\ - & - & - & - & - \\ - & - & - & - & - \\ - & - & - & - & ⑥ \\ - & ⑧ & ⑥ & - & - \end{bmatrix} \begin{array}{c} * \\ * \\ * \\ * \\ * \end{array}$$

Fifth iteration

STEP 2 No uncircled numbers exist in the starred rows, so we STOP.

The minimum connector has length $5 + 6 + 6 + 8 = 25$, and includes the edges AD, DE, EB and EC. This is the minimum connector we obtained in parts (1) and (2).

(b) (1) Removing town B, we obtain the following minimum spanning tree with total weight 17.

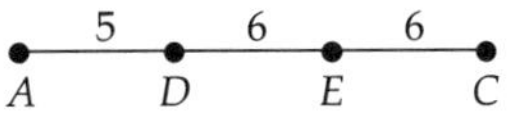

•——5——•——6——•——6——•
A   D   E   C

Adding the two smallest weights at B (8 and 9), we obtain the lower bound $17 + (8 + 9) = 34$.

(2) Removing town E, we obtain one of six minimum spanning trees with total weight 21.

Adding the two smallest weights at E (6 and 6), we obtain the lower bound $21 + (6 + 6) = 33$.

The correct solution is 35 — for example, $ADECBA$ or $ADBECA$.

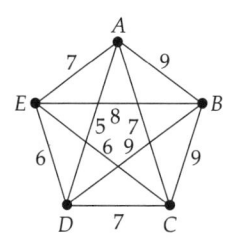

Each of the six spanning trees comprises the edge AD with weight 5, one of the edges of weight 7 (AC or CD), and one of the edges of weight 9 (AB or BC or BD).

3.5 For example, removing the vertex Berlin, we obtain either of the following minimum spanning trees with total weight 38.

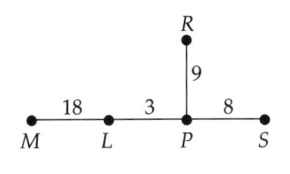

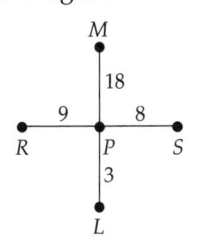

Adding the two smallest weights at Berlin (7 and 7), we obtain the lower bound $38 + (7 + 7) = 52$.

Similarly, by removing each of the other vertices, we obtain the following lower bounds:

London $35 + (3 + 7) = 45$;
Moscow $27 + (11 + 18) = 56$;
Paris $39 + (3 + 7) = 49$;
Rome $29 + (9 + 10) = 48$;
Seville $30 + (8 + 11) = 49$.

The larger of your two lower bounds is the better one.

4.1

(a) We find the following maximum flows by inspection:

$f(AB) = 5$, $f(AC) = 5$, $f(AD) = 5$, $f(BC) = 8$, $f(BD) = 7$, $f(CD) = 7$.

So the complete network $C(N)$ is:

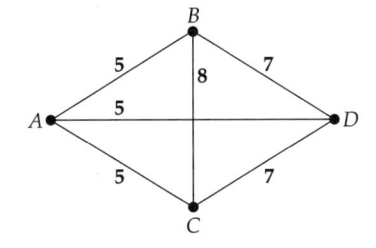

(1)

cycle	smallest capacity	number of occurrences
ABCA	5	2
ABDA	5	2
ACDA	5	2
BCDB	7	2
ABCDA	5	2
ABDCA	5	2
ADBCA	5	2

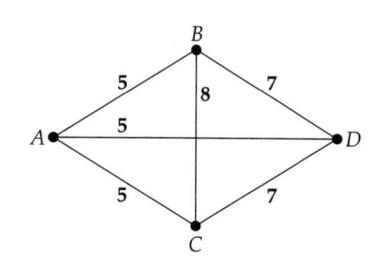

So the smallest capacity occurs more than once in each cycle.

(2) There are only three capacities in $C(N)$ — 5, 7 and 8.

(b) There are six maximum spanning trees. Any maximum spanning tree for this network must include the edge with capacity 8 (BC), one of the edges with capacity 7 (BD or CD, but not both, since this would create a cycle), and any one of the edges with capacity 5. One such tree is as shown in the margin.

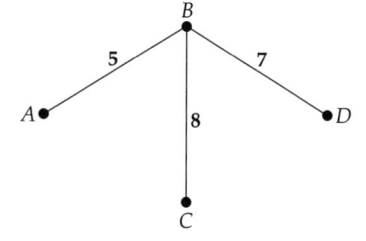

4.2 Using the capacity rule, we obtain the following table of maximum flows.

	A	B	C	D	E	F
A	–	2	2	2	2	1
B	2	–	4	3	3	1
C	2	4	–	3	3	1
D	2	3	3	–	5	1
E	2	3	3	5	–	1
F	1	1	1	1	1	–

4.3

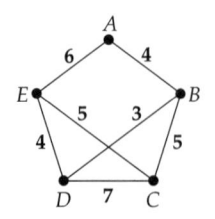

The given network has 5 vertices, so to find a cut tree we perform 4 iterations.

First iteration

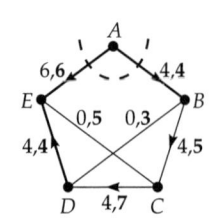

First, we find a maximum flow from A to E.

max flow from A to E has value 10;

min cut separates $\{A\}$, $\{B, C, D, E\}$;

draw branch: $A \overset{10}{\longrightarrow} (BCDE)$

60

Second iteration

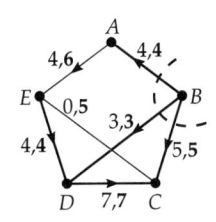

Next, we find a maximum flow from B to C.

max flow from B to C has value 12;

min cut separates $\{B\}, \{A, C, D, E\}$;

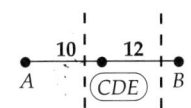

draw branch: $B \overset{12}{\longleftrightarrow} \boxed{CDE}$

join to: $A \overset{10}{\longleftrightarrow} \cdots$

join branch to tree: $A \overset{10}{\longleftrightarrow} \boxed{CDE} \overset{12}{\longleftrightarrow} B$

Third iteration

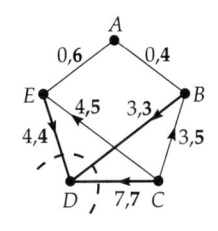

Next, we find a maximum flow from C to D.

max flow from C to D has value 14;

min cut separates $\{D\}, \{A, B, C, E\}$;

draw branch: $D \overset{14}{\longleftrightarrow} \boxed{CE}$

join to: $A \overset{10}{\longleftrightarrow} \cdots \overset{12}{\longleftrightarrow} B$

join branch to tree:

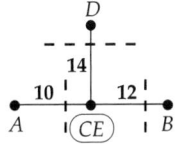

$$
\begin{array}{c}
D \\
| \\
14 \\
A \overset{10}{\longleftrightarrow} \boxed{CE} \overset{12}{\longleftrightarrow} B
\end{array}
$$

Fourth iteration

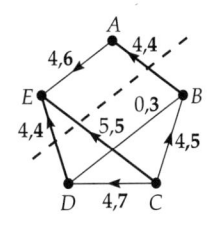

Finally, we find a maximum flow from C to E.

max flow from C to E has value 13;

min cut separates $\{A, E\}, \{B, C, D\}$;

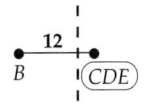

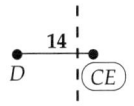

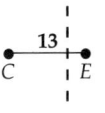

draw branch: $C \overset{13}{\longleftrightarrow} E$

join to:

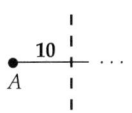

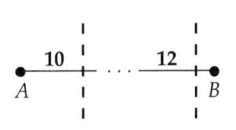

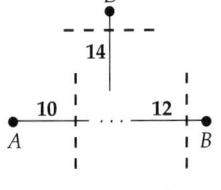

$$
\begin{array}{c}
D \\
| \\
14 \\
A \overset{10}{\longleftrightarrow} \cdots \overset{12}{\longleftrightarrow} B
\end{array}
$$

join branch to tree:

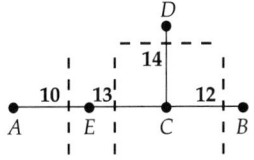

$$
\begin{array}{c}
D \\
| \\
14 \\
A \overset{10}{\longleftrightarrow} E \overset{13}{\longleftrightarrow} C \overset{12}{\longleftrightarrow} B
\end{array}
$$

The last tree is the required cut tree.

Using the capacity rule, we obtain the following table of maximum flows.

	A	B	C	D	E
A	–	10	10	10	10
B	10	–	12	12	12
C	10	12	–	14	13
D	10	12	14	–	13
E	10	12	13	13	–

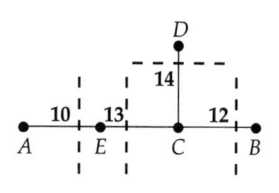

4.4

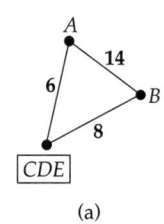

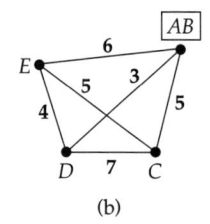

(a) (b)

4.5

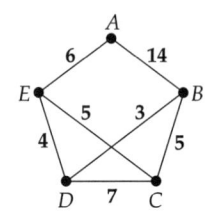

The given network has 5 vertices, so to find a cut tree we perform 4 iterations.

First iteration

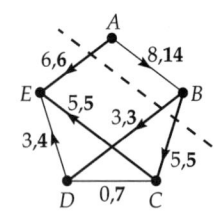

First, we find a maximum flow from A to E.

max flow from A to E has value 14;

min cut separates $\{A, B\}$, $\{C, D, E\}$;

draw branch:

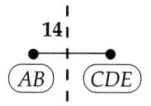

Second iteration

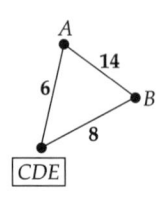

Next, we find a maximum flow from A to B. By the condensing theorem, we may condense the vertices C, D and E, as in Exercise 4.4(a).

max flow from A to B has value 20;

min cut separates $\{A\}$, $\{B, C, D, E\}$;

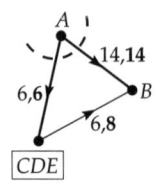

draw branch:

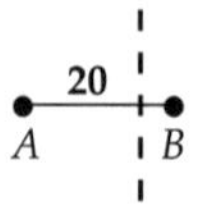

join branch to tree:

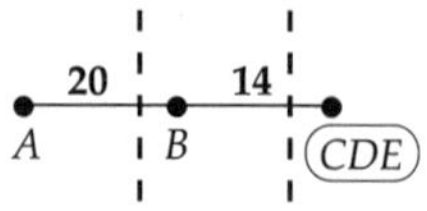

Third iteration

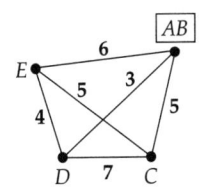

Next, we find a maximum flow from C to D. By the condensing theorem, we may condense the vertices A and B, as in Exercise 4.4(b).

max flow from C to D has value 14;

min cut separates $\{D\}, \{A, B, C, E\}$;

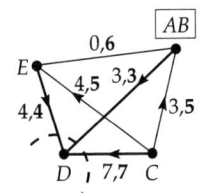

draw branch:

$$D \overset{14}{\longrightarrow} CE$$

join branch to tree:

$$A \overset{20}{\longrightarrow} B \overset{14}{\longrightarrow} CE \overset{14}{\longrightarrow} D$$

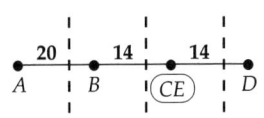

Fourth iteration

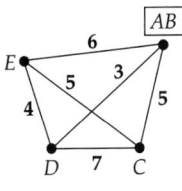

Finally, we find a maximum flow from C to E. Again, we may condense the vertices A and B.

max flow from C to E has value 15;

min cut separates $\{E\}, \{A, B, C, D\}$;

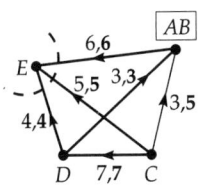

draw branch:

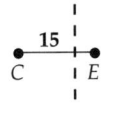

$$C \overset{15}{\longrightarrow} E$$

join branch to tree:

$$A \overset{20}{\longrightarrow} B \overset{14}{\longrightarrow} C \overset{14}{\longrightarrow} D, \quad C \overset{15}{\longrightarrow} E$$

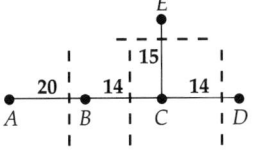

The last tree is the required cut tree.

Solutions to the problems

Solution 1.1

The six unlabelled trees with six vertices are as follows.

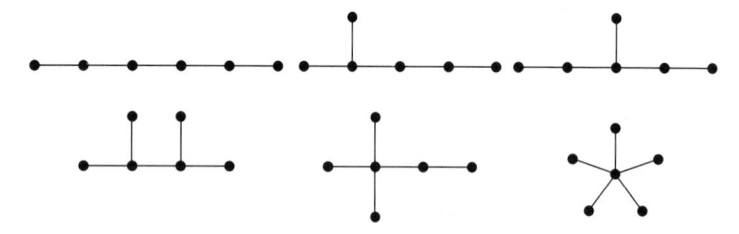

Solution 1.2

The eleven unlabelled trees with seven vertices are as follows.

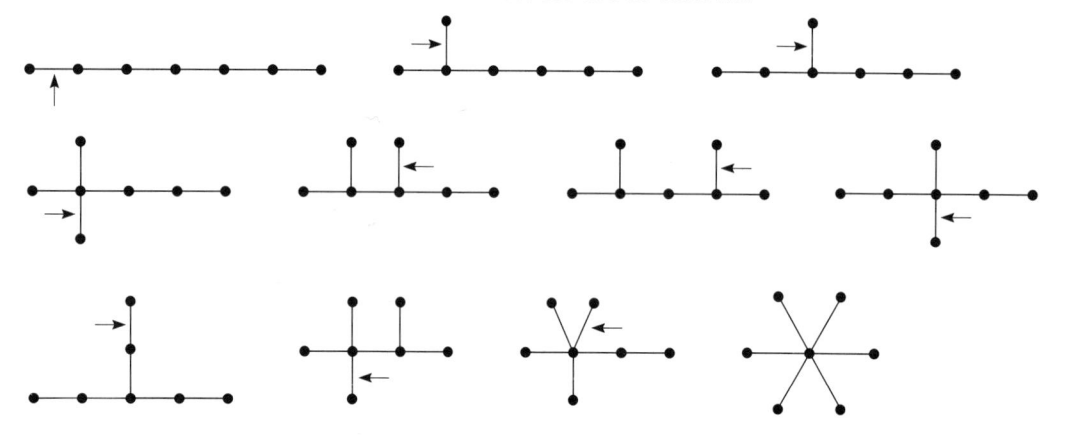

Solution 1.3

(a) Suppose that removing the edge e disconnects the graph into more than two components. Since e joins only two vertices, it can link at most two of these components, so that at least one component remains disconnected from the rest when e is put back into the tree. This contradicts the fact that the tree is connected. Thus, the removal of e disconnects the tree into just two components.

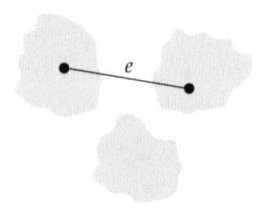

(b) Suppose that the addition of an edge e creates two or more cycles:

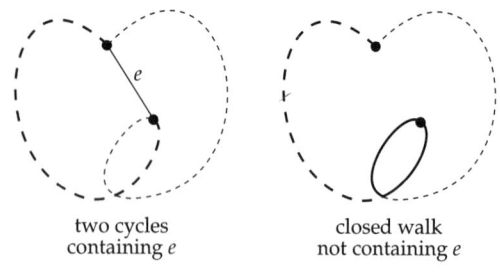

two cycles containing e

closed walk not containing e

The parts of any two such cycles other than e can be combined into a closed walk that does not contain e, and this closed walk must contain a cycle. This contradicts the fact that a tree contains no cycles. Thus, the addition of a new edge cannot create more than one cycle.

Solution 1.4

(a)

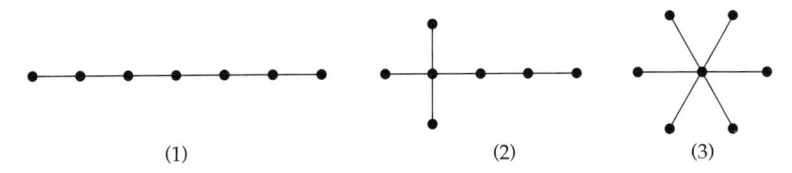

(1) (2) (3)

(b) Let T be a tree with n vertices and at most one vertex of degree 1. Then T has at least $n - 1$ vertices of degree 2 or more. It follows that the sum of the vertex degrees is at least $2(n - 1) + 1$, and so, by the handshaking lemma, the number of edges of T is at least $(n - 1) + \frac{1}{2}$. This contradicts the fact that T has exactly $n - 1$ edges, and the result follows immediately.

Solution 1.5

The tree has seven downward edges from each vertex:

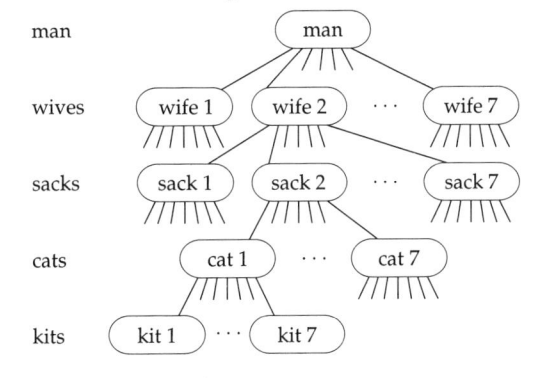

man
wives
sacks
cats
kits

Solution 1.6

In the branching tree representing the outcomes of two throws of a six-sided die, there are three 'levels' (including the root) with six downward edges from each vertex:

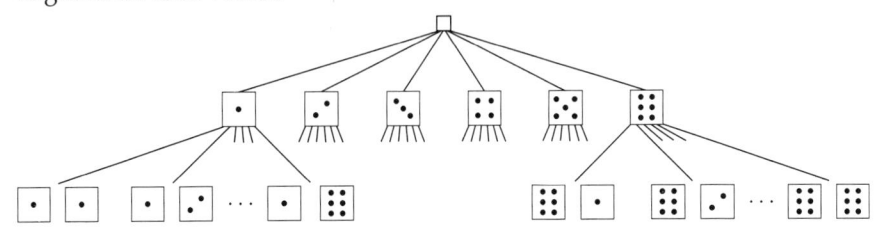

Solution 1.7

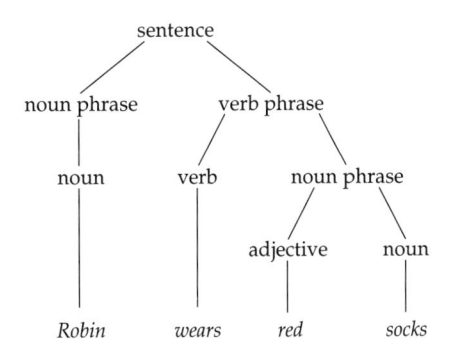

sentence — noun phrase — noun — Robin; verb phrase — verb — wears; noun phrase — adjective — red, noun — socks

Solution 1.8

The sorting of mail can be represented by a tree similar to that used in the Dewey decimal system example. A letter with postcode MK7 6AA (for example) is first sorted into the MK box, then the MK7 box, then the MK7 6 box, and so on.

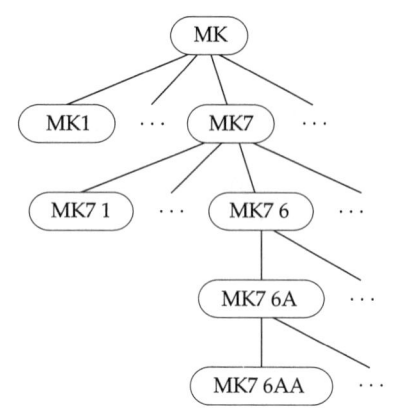

Solution 1.9

The subsets of a set corresponding to this tree can be drawn as follows.

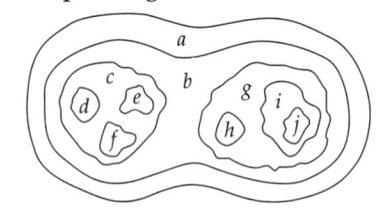

The nested parentheses corresponding to this tree are:

$$((((\ )(\ )(\ ))((\ )((\ )))))$$

Solution 2.1

The sixteen labelled trees with four vertices are as follows. The first four arise from labelling the complete bipartite graph $K_{1,3}$, and the others arise from labelling the path graph P_4.

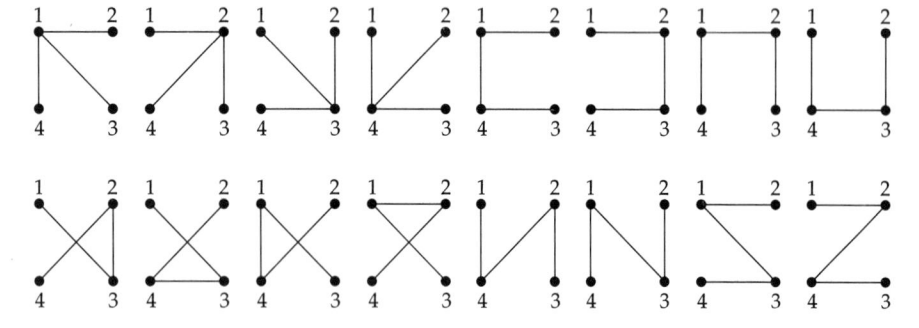

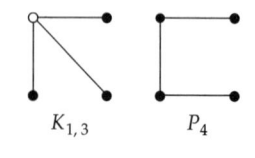

Solution 2.2

For each value of n, the number of labelled trees with n vertices is a power of n; it is, in fact, n^{n-2}.

Solution 2.3

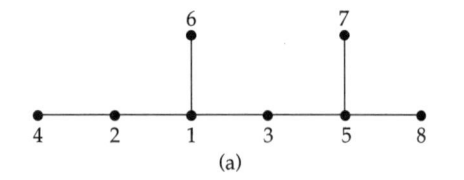

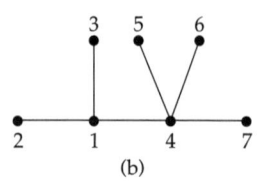

(a) Successively removing the edges 42, 21, 61, 13, 35 and 75, we obtain the Prüfer sequence (2, 1, 1, 3, 5, 5).

(b) Successively removing the edges 21, 31, 14, 54 and 64, we obtain the Prüfer sequence (**1, 1, 4, 4, 4**).

Solution 2.4

(a) We start with the list (1, 2, 3, 4, 5, 6, 7, 8) and the sequence (**2, 1, 1, 3, 5, 5**). Successively adding the edges 42, 21, 61, 13, 35 and 75 leaves us with the list (5, 8). Joining the vertices with these labels, we obtain the labelled tree (a) in Problem 2.3.

(b) We start with the list (1, 2, 3, 4, 5, 6, 7) and the sequence (**1, 1, 4, 4, 4**). Successively adding the edges 21, 31, 14, 54 and 64 leaves us with the list (4, 7). Joining the vertices with these labels, we obtain the labelled tree (b) in Problem 2.3.

Solution 2.5

The sixteen labelled trees with four vertices, and their associated Prüfer sequences are as follows.

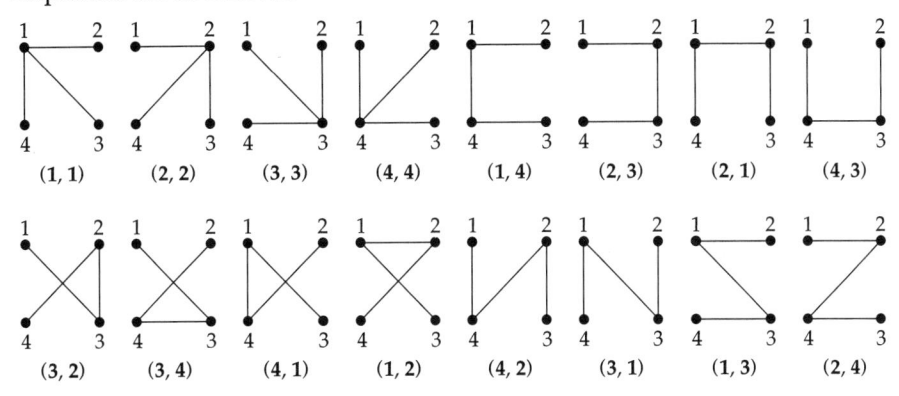

Note that each of the sixteen possible sequences occurs exactly once.

Solution 2.6

Each canal system corresponds to a labelled tree with eight vertices. By Cayley's theorem, there are $8^6 = 262\,144$ of these.

Solution 2.7

The fourteen binary trees with four vertices are:

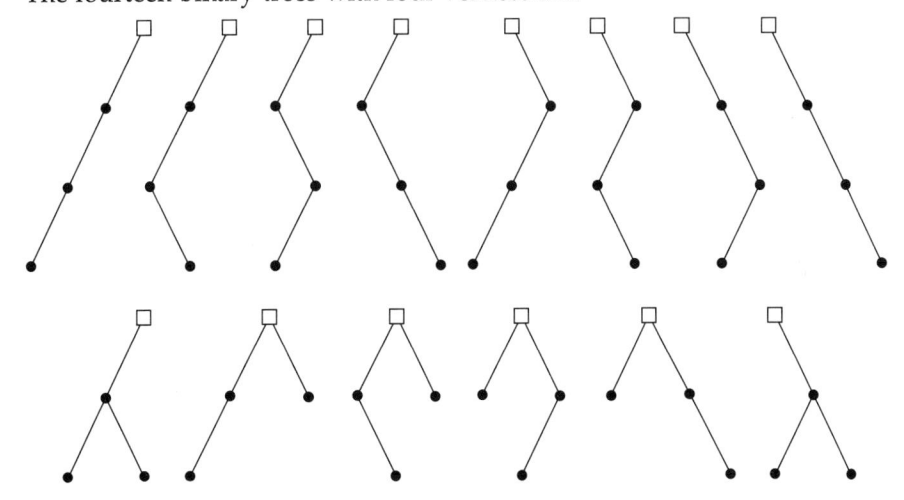

Solution 2.8

Substituting the value $n = 6$ in the recurrence relation, we obtain

$$u_6 = 2u_5 + (u_1 u_4 + u_2 u_3 + u_3 u_2 + u_4 u_1)$$
$$= (2 \times 42) + [(1 \times 14) + (2 \times 5) + (5 \times 2) + (14 \times 1)] = 132.$$

Thus there are 132 binary trees with six vertices.

Solution 2.9

The number of vertices in the graph of a molecule with formula C_6H_{14} is

$6 + 14 = 20$.

By the handshaking lemma, the number of edges is half the sum of the vertex degrees, that is,

$\frac{1}{2}[(6 \times 4) + (14 \times 1)] = 19$.

Since the graph is connected, and the number of vertices exceeds the number of edges by 1, the graph is a tree, by Theorem 1.1.

Solution 2.10

The number of vertices is $n + (2n + 2) = 3n + 2$.

By the handshaking lemma, the number of edges is

$\frac{1}{2}[(n \times 4) + ((2n + 2) \times 1)] = \frac{1}{2}(4n + (2n + 2)) = 3n + 1$.

Since the graph is connected, and the number of vertices exceeds the number of edges by 1, the graph is a tree, by Theorem 1.1.

Solution 2.11

central with centre v

bicentral with bicentre vw

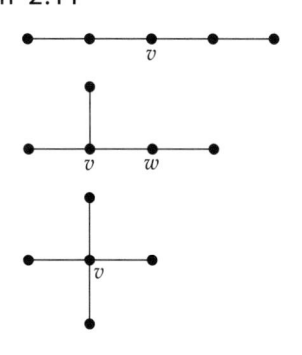

central with centre v

bicentral with bicentre vw

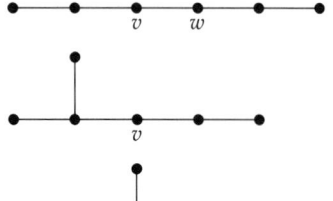

central with centre v

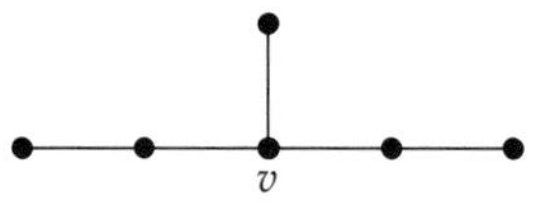

central with centre v

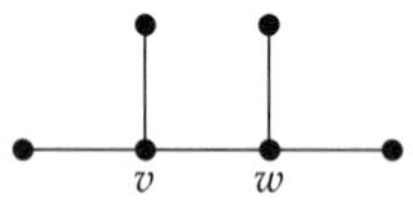

bicentral with bicentre vw

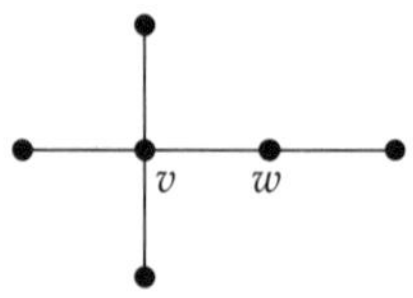

bicentral with bicentre vw

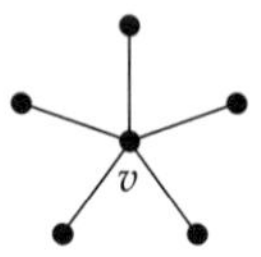

central with centre v

Solution 3.1

The eighteen spanning trees (other than those depicted in the text) are:

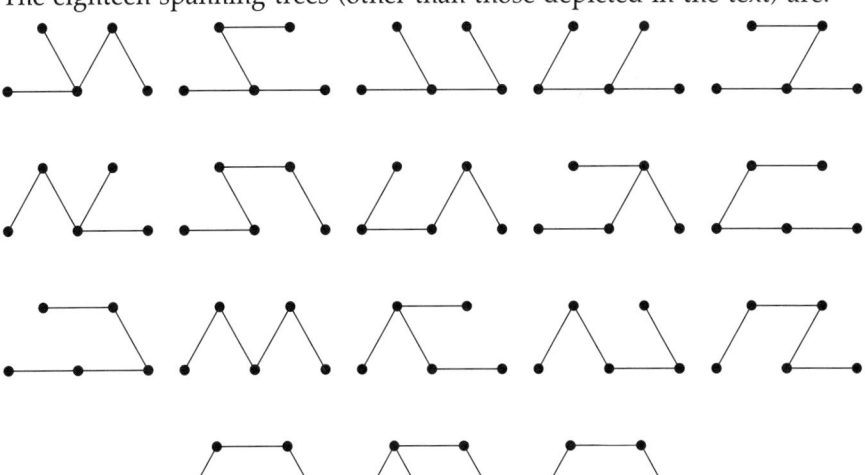

For clarity, labels have been omitted from these diagrams.

Solution 3.2

Three spanning trees of the Petersen graph are:

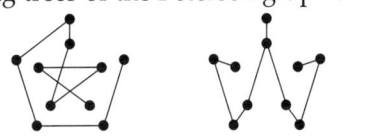

There are 1997 other possibilities, so we cannot show them all! The Petersen graph has 10 vertices, so a spanning tree is a connected graph with 9 edges linking all the vertices. Check that your trees have this property.

Solution 3.3

Each spanning tree in K_n corresponds to a labelled tree with n vertices. For example, each spanning tree in K_5 has five vertices:

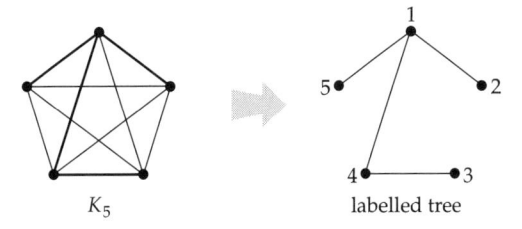

By Cayley's theorem, the number of spanning trees in K_n is n^{n-2}.

Solution 3.4

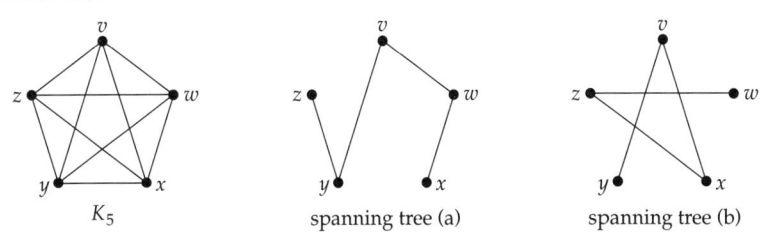

Building-up method: If we choose the edges vw, wx, vy and yz, then no cycles are created, and we obtain spanning tree (a) above.

Cutting-down method: If we remove the edges vw (destroying the cycle $vwxv$), wx (destroying the cycle $wxyw$), xy (destroying the cycle $xyzx$), yz (destroying the cycle $vyzv$), vz (destroying the cycle $vxzv$) and wy (destroying the cycle $vxzwyv$), then no cycles remain, and we obtain the spanning tree with edges vx, xz, wz and vy, that is, spanning tree (b) above.

The number of labelled spanning trees of K_5 is $5^{5-2} = 5^3 = 125$. (See Solution 3.3.)

Solution 3.5

We should have chosen edges in the order *AE* (length 2), *AC* (length 4), *BC* (length 5) and *DE* (length 7), obtaining the minimum spanning tree shown in the margin.

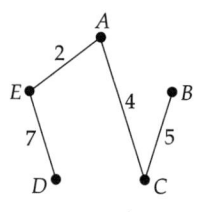

Solution 3.6

We apply Kruskal's algorithm as follows.

First edge   We choose Athlone–Galway (weight 56).

Second edge   We choose Galway–Limerick (weight 64).

Third edge   We choose Athlone–Sligo (weight 71).

Fourth edge   We cannot choose Athlone–Limerick (weight 73), as this creates a cycle, so we choose Athlone–Dublin (weight 78).

Fifth edge   We cannot choose Galway–Sligo (weight 85), as this creates a cycle, so we choose Dublin–Wexford (weight 96).

This completes the required minimum spanning tree of total weight 365.

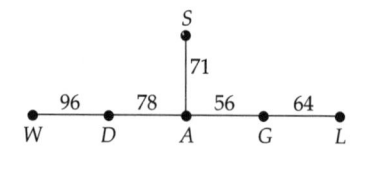

Solution 3.7

(a) We should have chosen edges in the order Berlin–Paris (weight 7), Paris–London (weight 3), Paris–Seville (weight 8), Paris–Rome (weight 9), Berlin–Moscow (weight 11), obtaining the spanning tree of total weight 38, shown in the margin.

(b) We should have chosen edges in the order Rome–Paris (weight 9), Paris–London (weight 3), Paris–Berlin or London–Berlin (weight 7), Paris–Seville (weight 8), Berlin–Moscow (weight 11), obtaining the same spanning tree as in part (a) of total weight 38 or the spanning tree obtained in Example 3.3 (depending on the choice of edge of weight 7).

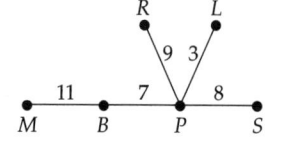

Solution 3.8

We start with the matrix:

$$\begin{array}{c} & \begin{array}{ccccc} 1 & 2 & 3 & 4 & 5 \end{array} \\ \begin{array}{c} 1 \\ 2 \\ 3 \\ 4 \\ 5 \end{array} & \left[\begin{array}{ccccc} - & 6 & 4 & 8 & 2 \\ 6 & - & 5 & 8 & 6 \\ 4 & 5 & - & 9 & 4 \\ 8 & 8 & 9 & - & 7 \\ 2 & 6 & 4 & 7 & - \end{array} \right] \end{array}$$

STEP 1 We delete all entries in column 1, and mark row 1 with a star.

$$\begin{array}{c} & \begin{array}{ccccc} 1 & 2 & 3 & 4 & 5 \end{array} \\ \begin{array}{c} 1 \\ 2 \\ 3 \\ 4 \\ 5 \end{array} & \left[\begin{array}{ccccc} - & 6 & 4 & 8 & 2 \\ - & - & 5 & 8 & 6 \\ - & 5 & - & 9 & 4 \\ - & 8 & 9 & - & 7 \\ - & 6 & 4 & 7 & - \end{array} \right]* \end{array}$$

First iteration

STEP 2 The smallest uncircled entry in the starred row is 2 in column 5, so we select 2.

STEP 3 We circle the 2 in row 1 and column 5, mark row 5 with a star, and delete the remaining entries in column 5.

$$\begin{array}{c} & \begin{array}{ccccc} 1 & 2 & 3 & 4 & 5 \end{array} \\ \begin{array}{c} 1 \\ 2 \\ 3 \\ 4 \\ 5 \end{array} & \left[\begin{array}{ccccc} - & 6 & 4 & 8 & ② \\ - & - & 5 & 8 & - \\ - & 5 & - & 9 & - \\ - & 8 & 9 & - & - \\ - & 6 & 4 & 7 & - \end{array} \right]\begin{array}{c}* \\ \\ \\ \\ *\end{array} \end{array}$$

70

Second iteration

STEP 2 The smallest uncircled entry in the starred rows is 4 in each starred row and column 3; let us select 4 in row 1 and column 3.

STEP 3 We circle 4 in row 1 and column 3, mark row 3 with a star, and delete the remaining entries in column 3.

$$\begin{array}{c} & 1 & 2 & 3 & 4 & 5 \\ 1 & \left[- \right. & 6 & ④ & 8 & ② \\ 2 & - & - & - & 8 & - \\ 3 & - & 5 & - & 9 & - \\ 4 & - & 8 & - & - & - \\ 5 & \left. - \right. & 6 & - & 7 & - \end{array} \begin{array}{c} * \\ \\ * \\ \\ * \end{array}$$

Third iteration

STEP 2 The smallest uncircled entry in the starred rows is 5 in row 3 and column 2, so we select 5.

STEP 3 We circle 5 in row 3 and column 2, mark row 2 with a star, and delete the remaining entries in column 2.

$$\begin{array}{c} & 1 & 2 & 3 & 4 & 5 \\ 1 & \left[- \right. & - & ④ & 8 & ② \\ 2 & - & - & - & 8 & - \\ 3 & - & ⑤ & - & 9 & - \\ 4 & - & - & - & - & - \\ 5 & \left. - \right. & - & - & 7 & - \end{array} \begin{array}{c} * \\ * \\ * \\ \\ * \end{array}$$

Fourth iteration

STEP 2 The smallest uncircled entry in the starred rows is 7 in row 5 and column 4, so we select 7.

STEP 3 We circle 7 in row 5 and column 4, mark row 4 with a star, and delete the remaining entries in column 4.

$$\begin{array}{c} & 1 & 2 & 3 & 4 & 5 \\ 1 & \left[- \right. & - & ④ & - & ② \\ 2 & - & - & - & - & - \\ 3 & - & ⑤ & - & - & - \\ 4 & - & - & - & - & - \\ 5 & \left. - \right. & - & - & ⑦ & - \end{array} \begin{array}{c} * \\ * \\ * \\ * \\ * \end{array}$$

Fifth iteration

STEP 2 No uncircled numbers exist in the starred rows, so we STOP.

The minimum connector has length 2 + 4 + 5 + 7 = 18, and includes the edges *AE*, *AC*, *CB* and *ED*.

Solution 3.9

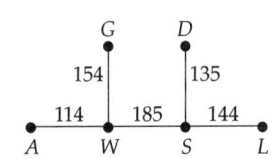

We apply Prim's algorithm for maximum spanning trees as follows.

First edge We choose Athlone–Wexford (weight 114).
Second edge We choose Wexford–Sligo (weight 185).
Third edge We choose Wexford–Galway (weight 154).
Fourth edge We choose Limerick–Sligo (weight 144).
Fifth edge We choose Dublin–Sligo (weight 135).

This completes the required maximum spanning tree of total weight 732.

Solution 3.10

(a)

total weight ≥ (7 + 2 + 4) + (6 + 5) = 24

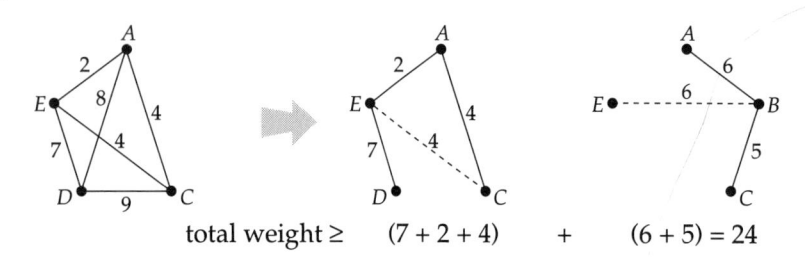

The minimum spanning tree joining the vertices A, C, D and E is the tree with edges AE, DE, and AC or CE, with total weight 13. The two edges of smallest weight incident to B are BC, and BA or BE, with total weight 11. The lower bound is therefore $13 + 11 = 24$.

(b)

total weight ≥ (8 + 4 + 5) + (2 + 4) = 23

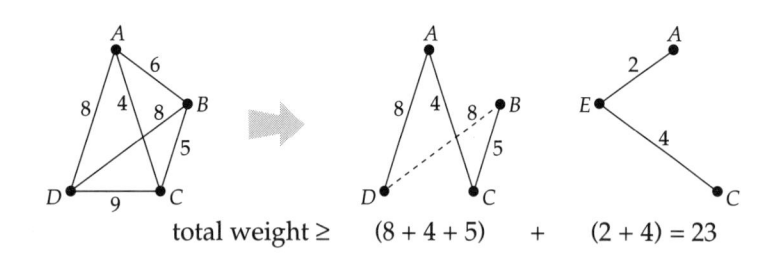

The minimum spanning tree joining the vertices A, B, C and D is the tree with edges AC, BC, and AD or BD, with total weight 17. The two edges of smallest weight incident to E are EA and EC, with total weight 6. The lower bound is therefore $17 + 6 = 23$.

The better lower bound is that given by part (a), that is, 24.

Solution 3.11

```
   96    116    64    85        71    56
●─────●─────●─────●─────●     ●─────●─────●
 D     W     L     G     S     S     A     G
```

total weight ≥ (96 + 116 + 64 + 85) + (71 + 56) = 488

The minimum spanning tree joining the vertices Dublin, Galway, Limerick, Sligo and Wexford is the tree with edges Galway–Limerick, Galway–Sligo, Dublin–Wexford and Limerick–Wexford, with total weight 361.

The two edges of smallest weight incident to Athlone are Athlone–Galway and Athlone–Sligo, with total weight 127. The lower bound is therefore $361 + 127 = 488$.

(The actual solution is obtained by visiting the places in the order Athlone → Sligo → Galway → Limerick → Wexford → Dublin → Athlone, or vice versa, covering a distance of 510 miles. The lower bound of 488 is therefore correct to within 5%.)

Solution 4.1

(a) $f(AB) = 4$, $f(AC) = 9$, $f(AD) = 9$, $f(AE) = 2$, $f(BC) = 4$, $f(BD) = 4$, $f(BE) = 2$, $f(CD) = 10$, $f(CE) = 2$, $f(DE) = 2$.

(b) The complete network $C(N)$ is shown in the margin.

1. In each cycle of $C(N)$, the smallest capacity occurs more than once; for example, in the cycle $ABCDA$, the capacity 4 occurs twice.

2. $C(N)$ has five vertices, but only four different capacities: 2, 4, 9 and 10.

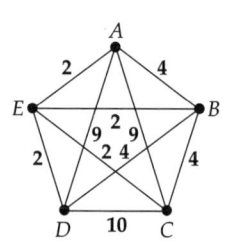

Solution 4.2

By the capacity rule, we have

$f(AD) = \min(f(AB), f(BD)) = \min(8, 5) = 5$,

$f(BC) = \min(f(BA), f(AC)) = \min(8, 7) = 7$,

$f(CD) = \min(f(CA), f(AB), f(BD)) = \min(7, 8, 5) = 5$.

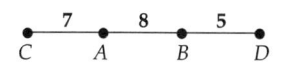

Solution 4.3

By the capacity rule, we have

$f(AC) = \min(f(AB), f(BC)) = \min(2, 7) = 2$,

$f(AD) = \min(f(AB), f(BD)) = \min(2, 3) = 2$,

$f(AE) = \min(f(AB), f(BD), f(DE)) = \min(2, 3, 5) = 2$,

$f(BE) = \min(f(BD), f(DE)) = \min(3, 5) = 3$,

$f(CD) = \min(f(CB), f(BD)) = \min(7, 3) = 3$,

$f(CE) = \min(f(CB), f(BD), f(DE)) = \min(7, 3, 5) = 3$.

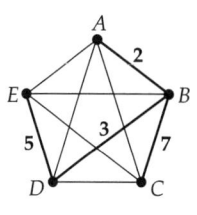

Solution 4.4

Let $V'W'$ be an edge of smallest capacity in the path joining V and W in the maximum spanning tree. If we now add the edge VW to the maximum spanning tree, we get a cycle containing both VW and $V'W'$. By applying Theorem 4.1 to this cycle, we see that the capacity of VW cannot be smaller than the capacity of $V'W'$. But the capacity of VW cannot be *larger* than the capacity of $V'W'$ because if it were we could replace $V'W'$ by VW; this would give us a spanning tree of greater total capacity than our original maximum spanning tree, which is impossible. It follows that $f(VW) = f(V'W')$, as required.

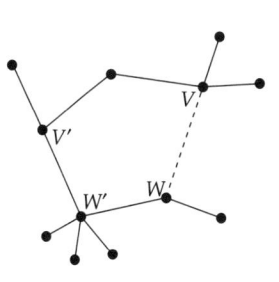

Solution 4.5

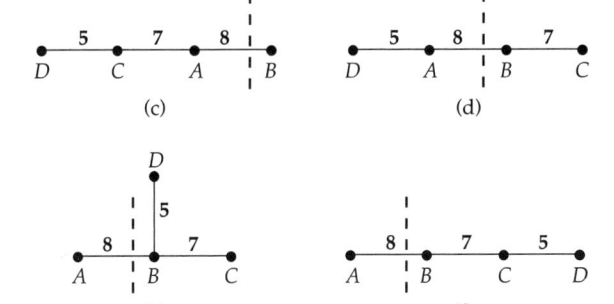

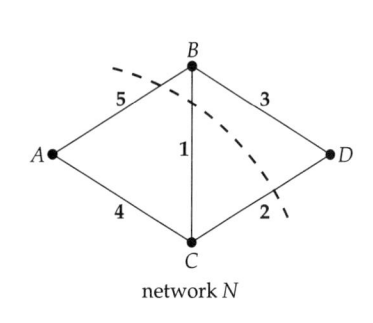

The minimum cuts for the maximum flow from A to B are:

tree (c) $\{A, C, D\}, \{B\}$; tree (d) $\{A, D\}, \{B, C\}$;
tree (e) $\{A\}, \{B, C, D\}$; tree (f) $\{A\}, \{B, C, D\}$.

However, a maximum flow from A to B in the network N corresponds to the minimum cut separating the vertices into the two sets $\{A, C\}$ and $\{B, D\}$, so none of the above trees is a cut tree for N.

Solution 4.6

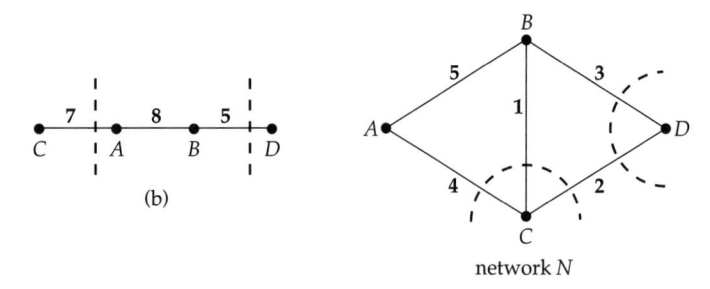

(b) diagram: C —7— A —8— B —5— D (with cut lines)

network N: A-B (5), B-D (3), A-C (4), C-D (2), B-C (1)

(a) In both the network and the tree, a minimum cut of capacity 5 separates the vertices into two sets {A, B, C} and {D}.

(b) In both the network and the tree, a minimum cut of capacity 7 separates the vertices into two sets {A, B, D} and {C}.

Index

algorithms
   Gomory and Hu's 45
   Kruskal's greedy 28, 33, 41
   Prim's greedy 30, 31, 33, 35, 41
alkane 23

bicentral tree 24
bicentre (of tree) 24
bin-packing problem 49
binary tree 8, 21
bound, lower 36
branching tree 9
building-up method for constructing spanning tree 27

cable television network 48
capacity rule 42
Cayley's theorem 16, 20
Cayley, A. 21, 23, 24, 25
central tree 24
centre (of tree) 24
chemical tree 23
combinatorial explosion 16
complete network 40
complexity theory 49
conceptual tree 8
condensing vertices 45
Cook, S. 50
counting trees
   binary 21
   chemical 23
   labelled 16
   unlabelled 16
cut tree 45
cutting-down method for constructing spanning tree 27
cycle
   minimum-weight Hamiltonian 36

efficiency of algorithm 49
exponential-time algorithm 49

flow-equivalent networks 43
forest 52
   spanning 52

gas pipeline distribution system 7, 48
Gomory, R. E. 39
grammatical tree 10

greedy algorithm 28, 48
   Kruskal's 28, 33, 41
   Prim's 30, 31, 33, 35, 41

Hamiltonian cycle 36
   minimum-weight 36
Harary, F. 25
Hu, T. C. 39

knapsack problem 49
Kruskal's greedy algorithm 28, 33
   for maximum connector 41
Kruskal, J. B. 28

lower bound for the travelling salesman problem 36

max-flow min-cut theorem 40
maximum connector 34
   Kruskal's greedy algorithm for 41
   Prim's greedy algorithm for 34
maximum spanning tree 34, 41
minimax problem 50
minimum connector 27
   Kruskal's greedy algorithm for 28, 33
   Prim's greedy algorithm for 30, 31, 33
minimum connector problem 27
   extra locations for 46
minimum cut in undirected network 40
minimum spanning tree 27
minimum-weight Hamiltonian cycle 36
modelling
   with semi-lattices 13
   with trees 13
multi-terminal flow problem 39

network
   complete 40
   undirected 39, 40
NP-complete problem 36, 50

physical tree 7
Pólya, G. 25
polynomial-time algorithm 30, 49
Prim's greedy algorithm 30, 33
   for maximum connector 34
   tabular (matrix) form 31

problem
   easy 49
   hard 49
   minimax 51
   NP-complete 36, 50
Prüfer sequence 16
Prüfer, H. 16

Read, R. C. 25
recurrence relation 21
Redfield, H. 25
root (of tree) 8
rooted tree 8

semi-lattice 15
sorting tree 11
spanning
   forest 52
   tree 26, 47
      maximum 34, 41
      minimum 27, 28
stack 11
Steiner tree 47
Steiner, J. 47

telephone network 48
theorems
   1.1 equivalent definitions of tree 6
   2.1 Cayley's 20
   3.1 Prim's and Kruskal's algorithms 33
   4.1 smallest capacity in $C(N)$ 41
   4.2 different capacities in $C(N)$ 43
   condensing 45
tiling 50
travelling salesman problem 36, 49
tree 4
   bicentral 24
   binary 8, 21
   branching 8
   central 24
   chemical 23
   conceptual 8
   cut 45
   grammatical 10
   maximum spanning 34, 41
   minimum spanning 28
   physical 7
   rooted 8
   sorting 11
   spanning 26
   Steiner 47
tree, equivalent definitions of 6

undirected network 39, 40